THE MSP AI PLAYBOOK SERIES · BOOK 2

AI FOR MSPs

Managed Intelligence

The MSP Agent Operations Playbook

How to Build, Deploy, and Monetize AI Agent Management as the Next Evolution of Your MSP

Tony Ferrigno

Founder and Managing Partner, AiT Advisory Group

35+ Years of MSP Executive Leadership

Copyright Notice

AI for MSPs: Managed Intelligence

For permission requests, contact: tonyf@aitadvisorygroup.com or via aitadvisorygroup.com

Print ISBN: 979-8-950767-02-9

The information in this book is provided for general informational and educational purposes only. It does not constitute professional, legal, or financial advice. Results may vary based on your specific business circumstances. The author makes no representations or warranties regarding the accuracy or completeness of the information contained herein.

First Edition. Published 2026.

Printed in the United States of America.

Series Note

This is Book 2 in the series AI for MSPs: Managed Intelligence

Book 1: AI for MSPs: The No-Fluff Playbook.

Book 3: AI for MSPs: Process Intelligence and AI Readiness: The MSP Assessment Playbook.

AiT Advisory Group - We Put AI in IT

Table of Contents

Foreword: What Changes When You Stop Selling Tools and Start Operating Intelligence

> *"The MSPs who own the next decade are not the ones who sell the most AI tools. They are the ones who operate AI on behalf of their clients -- professionally, continuously, and with accountability for results."*
> **-- Tony Ferrigno**

I have been in managed services for more than 35 years. I have watched this industry navigate every major technology shift from break-fix to managed services, from on-premise to cloud, from reactive helpdesk to proactive NOC. Every single transition created winners and losers. The pattern never changed. The MSPs who moved early, built a real practice, and committed to the new model won. The ones who watched from the sidelines lost ground they never recovered.

AI agent management is that transition right now. And it is moving faster than anything I have seen before.

I wrote this book because Managed Intelligence is the most significant commercial opportunity in the history of the MSP industry. It is not a new product category or a vendor program. It is a new type of relationship between an MSP and its clients, one where your firm is the active operator of the AI systems running inside your clients' businesses. That relationship is stickier, higher-value, and harder to displace than anything traditional managed services has ever created.

The first two books in this series gave you the market case and the assessment methodology. This book gives you the delivery model, the governance framework, the pricing structure, and the organizational blueprint to build a Managed Intelligence practice that generates real recurring revenue and creates real competitive advantage.

Read every chapter. Work through the action items with your team. Then build, because the window for early-mover advantage is open right now, and it will not stay open.

Tony Ferrigno

Founder and Managing Partner, AiT Advisory Group

We Put AI in IT

INTRODUCTION

Managed Intelligence: What It Is, Who It Is For, and Why Now

Before anything else, you need a clear definition. Too many MSPs are using the phrase "AI services" to mean a dozen different things -- selling licenses, running training sessions, building chatbots, or deploying automation scripts. Managed Intelligence is something specific. Here is what it means.

What Managed Intelligence Is

Managed Intelligence is a purpose-built, ongoing managed service in which your MSP actively operates, monitors, governs, and optimizes your clients' AI agent infrastructure on their behalf -- with the same professionalism, rigor, and accountability you apply to managing their networks, security, and cloud environments.

It is not a product you resell. It is not a project you complete and hand off. It is a repeatable, structured managed service with defined service tiers, governance controls, performance reporting, and recurring revenue -- built around the operational lifecycle of AI agents running inside your clients' businesses.

The client gets the benefits of AI automation without the burden of managing it. You get high-margin MRR, deep client dependency, and a service category that very few of your competitors can match.

Who Managed Intelligence Is For

Managed Intelligence is designed for MSPs who are already delivering managed services to small and mid-size businesses and want to build a structured, monetizable AI practice on top of those existing relationships. You do not need to be a technology startup or a specialized AI firm. You need client relationships, operational discipline, and the willingness to learn a new delivery model.

Your existing clients are the market. They trust you. You have access to their systems. You understand how their businesses operate. That gives you a significant advantage over any new AI firm trying to enter the SMB market without those relationships.

If you are an MSP owner or executive who has been watching the AI conversation from the sidelines, unsure where to start and skeptical of the hype, this book is written specifically for you. It focuses on what is real, what is profitable, and what you can actually build.

Who Managed Intelligence Is Not For

This practice is not for MSPs looking for a quick add-on to sell in the next 30 days. Building a real Managed Intelligence practice takes 90 to 180 days of focused effort before the first client is fully operational. If you are not willing to invest in the methodology, the tooling, and the team training required to deliver this service with excellence, you will underdeliver and damage the client relationships you already have.

Done right, Managed Intelligence generates extraordinary recurring revenue and creates the kind of client relationships that competitors cannot easily displace. Done halfway, it creates expensive problems.

Why Now

The AI agent market crossed $5.4 billion in 2024 and is projected to reach $47.1 billion by 2030. That is a 44% compound annual growth rate. Seventy-two percent of Fortune 500 companies have active AI agent initiatives underway. The SMB market typically follows the enterprise curve with an 18-month lag. That lag is your window.

Your clients do not yet have anyone managing their AI agents. Most of them do not yet have AI agents in production. They are at the start of this adoption curve, looking to their technology partner for guidance. The MSPs who show up with a real service offering in 2026 will own this category in their markets before competitors understand what is happening.

That is the opportunity. This book shows you how to build the practice that captures it.

PART ONE

The Agent Economy

What Is Happening, Why It Matters, and What It Means for Your Business

CHAPTER ONE

Why AI Agents Are the Biggest Revenue Opportunity in MSP History

The Numbers, the Timing, and the Structural Advantage You Already Have

Start with this: the global AI agent market will grow from $5.4 billion in 2024 to $47.1 billion by 2030. That is not a forecast from a vendor with something to sell. That is the consensus view of multiple independent research organizations tracking enterprise AI adoption. The growth rate is 44% per year compounded.

For context, the cloud computing market grew at 23% per year from 2015 to 2020. That growth cycle created hundreds of millions of dollars in new recurring revenue for MSPs who built cloud practices early. AI agent management is growing at nearly twice that rate. The revenue opportunity is proportionally larger.

The SMB market is where you compete. And right now, your SMB clients are behind the enterprise curve by 18 months. That means the adoption surge in your market has not happened yet. You have time to build the practice, develop the methodology, and establish your reputation as the Managed Intelligence provider in your market before demand outpaces supply.

The Revenue Case

Here is what Managed Intelligence looks like financially for a single mid-size client with 50 users.

Revenue Component	Monthly Impact
Standard managed IT at $75 per user per month	$3,750 MRR
AI-enhanced managed IT at a 25% premium	$4,688 MRR
Managed Intelligence, three agents at mid-tier pricing	$3,200 MRR
Expansion agents added in months 7 through 12	$800 MRR growth
Total MRR at month 12, same client	$8,688
Revenue increase year one, same client	$59,256

That is not a best case. That is what happens when you add Managed Intelligence to a mid-size client who already pays you for managed services. The client gets measurably better business outcomes. You get $59,000 more in revenue from an account you already own. Run that math across ten clients and you have added nearly $600,000 in annual recurring revenue without signing a single new account.

The Structural Advantage You Already Have

Every previous technology wave in managed services rewarded scale and technical specialization. Cloud required infrastructure investment and vendor certifications. Cybersecurity required SOC infrastructure and threat intelligence capabilities. Those waves favored larger providers who could afford the upfront investment.

AI agent management is different. The primary competitive advantage in Managed Intelligence is not technical depth. It is operational intelligence about your clients' businesses. The MSP that wins in this category is the one who understands the client's processes well enough to know which ones should be automated, who knows how to design governance controls that satisfy compliance requirements, and who can demonstrate business impact month after month through consistent monitoring and reporting.

You have already built that intelligence. Years of managing your clients' technology environments has given you knowledge of how their businesses operate that no new AI firm entering the market can replicate quickly. Your existing client relationships are the moat. Managed Intelligence is how you monetize it.

The Competitive Window

Fewer than 18% of MSPs currently have a defined AI service offering of any kind. Of those, almost none have a structured, repeatable Managed Intelligence delivery model. The market is effectively open.

That will change. By 2027, AI agent management will be a standard expectation for MSPs competing for mid-market clients. The providers who build their practices in 2026 will have 24 months of delivery experience, case studies, client references, and operational efficiency that late entrants simply cannot match on a compressed timeline.

You do not have to be first to market in the entire industry. You have to be first in your market. The MSP that owns Managed Intelligence in Tampa, or Austin, or Charlotte, or wherever you operate has a defensible position that compounds over time.

Chapter 1 Action Items

1. Calculate your Managed Intelligence revenue potential. Identify your top 10 clients by revenue. For each, estimate how many AI agents could realistically be deployed in 18 months. Apply a conservative $800 per agent per month management fee. What does that number look like?
2. Identify your first prospect. Who in your portfolio is the most operationally mature, has the most manual process volume, and has the strongest relationship with your account team? This is where you start.
3. Commit to a practice build timeline. Set a specific date -- 90 days from today -- by which you will have your first Managed Intelligence client in production. Work backward from that date to define what you need to build.

CHAPTER TWO

How AI Agents Work

What You Need to Know Without the PhD

You do not need to be an AI engineer to run a Managed Intelligence practice. You need to understand how agents work well enough to scope engagements accurately, set client expectations honestly, and make good decisions about which tools to use for which problems. This chapter gives you that understanding without the academic theory.

The Plain Definition

An AI agent is a software system that uses artificial intelligence to pursue a specific goal through a sequence of autonomous steps. It perceives inputs from connected systems, makes decisions based on its instructions and reasoning capability, takes action in those systems, and adapts based on what happens. It does this continuously and without requiring a human to direct each individual step.

That is the key distinction. Traditional automation executes a fixed sequence of steps. An AI agent reasons through what to do next based on what it is seeing. When a helpdesk ticket arrives, a traditional automation script checks a predefined rule and routes it. An AI helpdesk agent reads the ticket, understands the context, considers the client's history, determines whether it can resolve the issue autonomously or needs to escalate, drafts a response if appropriate, and updates the ticket record. All of this happens in seconds, without a technician involved.

The Four-Step Agent Loop

Perceive: The agent receives information from its environment. An email arrives. A ticket is created. A form is submitted. A data anomaly is detected. The agent reads and understands the input, including its context.

Reason: The agent applies its instructions and reasoning capability to determine what action is appropriate. It considers the input, the applicable rules, the available options, and any relevant history.

Act: The agent takes action in the connected systems. It sends a response, creates a record, updates a status, routes a request, calls an API, or escalates to a human. These actions happen in real business systems your clients use every day.

Learn: Over time, well-designed agents improve. They identify patterns in which actions produce the best outcomes. They incorporate feedback from human reviewers. This continuous improvement is what justifies the ongoing management relationship and the ongoing monthly fee.

The Three Agent Types

Not all agents are equally complex or equally valuable. Understanding the three types helps you scope correctly and price appropriately.

Single-Task Agent	Multi-Step Workflow Agent	Orchestration Agent
Performs one specific task in response to a defined trigger. The logic is simple. The integration is usually one or two systems.	Executes a multi-step process involving multiple decisions, system interactions, and conditional branches.	Manages an entire business process, coordinates multiple sub-agents, and adapts in real time based on changing conditions.
Email triage, ticket routing, meeting scheduling, simple data lookup	Client onboarding, invoice processing, compliance reporting, sales qualification sequence	Full service desk autonomy, end-to-end procurement management, multi-system workflow coordination
Lowest complexity. Fastest to deploy. Easiest to govern.	Moderate complexity. 2 to 6 week deployment. Requires structured governance.	High complexity. 8 to 16 week deployment. Rigorous governance essential.
$300 to $800 per agent per month	$800 to $2,500 per agent per month	$2,500 to $6,000 per agent per month

The Platforms You Need to Know

The AI agent platform market is large and growing. For MSPs, four categories matter most right now.

Microsoft Copilot Studio

This is your starting point if you manage Microsoft 365 environments, which describes most MSPs. Copilot Studio is a low-code agent builder integrated natively with Microsoft 365, Teams, SharePoint, Dynamics, Power Platform, and Azure. Your technically capable team members can build, deploy, and manage sophisticated agents using a visual interface without writing custom code. For any MSP already using Microsoft 365, Copilot Studio is where you start building.

OpenAI Agents SDK

For more technically capable teams, the OpenAI Agents SDK enables highly capable custom agents that connect to virtually any system through API. The learning curve is steeper, but the ceiling on capability is significantly higher. This is the platform for complex custom agent work once your team has developed core competency.

Make and Zapier AI

Make (formerly Integromat) and Zapier with AI actions represent the bridge between traditional workflow automation and AI agency. These platforms allow agents to be built visually, with AI decision-making at key steps. If your team is stronger in automation than AI, these platforms offer the fastest path to initial deployments.

Vertical Platform Agents

ServiceNow, Salesforce Agentforce, HubSpot AI, and dozens of other business platforms are embedding agent capabilities directly into their products. Understanding which platforms your clients use and what agent capabilities are already available inside those platforms is a competitive advantage in client conversations and a shortcut to faster deployments.

Pro Tip

Build a platform competency map for your team before your first client conversation. Rate your current capability on each platform from 1 to 5. Identify the training required to reach a 4 on your primary platform. Set a timeline. Your first client engagement should use the platform your team is most ready to deliver on, not the most technically impressive option.

Why Agents Fail and Why That Creates Your Value

Unmanaged agents fail in specific, predictable ways. Understanding these failure modes is essential because they are exactly what your Managed Intelligence service prevents.

* Integration drift: The business systems agents connect to change their APIs, authentication tokens expire, or data schemas are updated. The agent loses connectivity and stops working. No one notices until a client complains.
* Behavioral drift: As the underlying AI model is updated by the platform vendor, agent behavior changes subtly. Outputs that were accurate six months ago become less accurate without anyone explicitly changing anything.
* Scope creep: Agents encounter situations outside their defined parameters and make plausible-sounding decisions that fall outside the intended scope. Without governance guardrails, these decisions go undetected.
* Data quality degradation: The data sources feeding the agent accumulate errors, duplicates, or outdated records. Agent performance degrades proportionally.

Every one of these failures is preventable through active monitoring and governance. And every one of them will happen to agents that are deployed and left unmanaged. That is not a risk you are protecting clients against in theory. That is a certainty you are protecting them against in practice.

Chapter 2 Action Items

1. Complete the Microsoft Copilot Studio learning path on Microsoft Learn. It is free and self-paced. Budget 20 hours for your primary team members to complete the core modules before your first client conversation.
2. Build one internal agent this month. Pick a high-frequency internal process -- scheduling, ticket triage, or proposal drafting. Build the agent, run it for 30 days, and measure the performance. Your first-hand experience is your most credible sales tool.
3. Identify the AI agent capabilities already embedded in the platforms your top 10 clients use. Most of them already have agent functionality available. This map tells you where your first Managed Intelligence conversations should happen.

CHAPTER TWO-B

Microsoft Agent 365: The MSP Control Plane for the Agentic Era

What It Is, How It Works, and Why MSPs Are the Natural Operators

Every major platform shift in technology eventually produces a control layer: a purpose-built management plane that sits above the individual tools and provides unified visibility, governance, and operational oversight. For cloud infrastructure, that layer is Azure or AWS management consoles. For endpoint management, it is your RMM. For the agentic era, Microsoft has built that layer, and it is called Agent 365.

Agent 365 is not a replacement for Copilot Studio or the other platforms where you build and configure AI agents. It is the operational layer that sits above them. Where Copilot Studio is the workshop where agents are constructed, Agent 365 is the control room where they are operated at scale. Understanding this distinction is essential to positioning your Managed Intelligence service, because Agent 365 is precisely the kind of platform that MSPs are built to manage and that most SMB clients are completely unequipped to run themselves.

What Agent 365 Is

Microsoft Agent 365 is an enterprise-grade agent management platform that provides centralized discovery, orchestration, governance, observability, security enforcement, and lifecycle management for AI agents running across Microsoft 365, Azure, Dynamics 365, and connected third-party environments. It is available through the Microsoft 365 admin center and accessible to organizations through Microsoft 365 enterprise licensing tiers.

Think of Agent 365 as the management infrastructure for your client's entire agent estate. Rather than managing each agent individually through the platform where it was built, Agent 365 gives you a single pane of glass across every agent the client has deployed, regardless of where it was built or which Microsoft surface it operates in. For an MSP managing agent estates across dozens of clients, this unified visibility is not a convenience feature. It is an operational necessity.

The Six Core Functions

Agent 365 performs six distinct functions that map directly to the Monitor and Govern pillars of your Managed Intelligence framework.

Function 1: Agent Discovery and Inventory. Before you can manage an agent estate, you need to know what is in it. Agent 365 automatically discovers and catalogs every AI agent operating within the tenant, including agents built in Copilot Studio, agents embedded in Microsoft 365 applications, agents created by end users through Microsoft Copilot, and agents connected through the Azure AI Foundry. The inventory includes the agent's name, creator, deployment date, connected systems, permission scope, and activity status. For MSPs taking over or auditing a client's existing agent environment, this discovery capability alone is worth the platform access.

Function 2: Centralized Orchestration. Agent 365 provides a unified interface for managing agent behavior across the tenant. You can configure which agents are active, what permissions they hold, what data they can access, and what actions they are authorized to take, without navigating to each individual agent's build environment. For clients with three or more deployed agents, centralized orchestration reduces the operational overhead of agent management by 40 to 60 percent compared to managing agents platform by platform.

Function 3: Governance Policy Enforcement. This is the function that directly strengthens your Govern pillar. Agent 365 enforces governance policies at the tenant level, not just at the individual agent level. You can define organization-wide rules governing what agents can access, what actions they can take autonomously, what data classifications they are permitted to process, and what external systems they can connect to. These tenant-level controls create a governance floor that applies to every agent in the environment, even newly created ones. In regulated industries, this floor is the difference between a defensible AI governance posture and an audit liability.

Function 4: Observability and Reporting. Agent 365 provides real-time telemetry on every deployed agent: activity logs, action histories, performance metrics, and anomaly flags. This telemetry feeds directly into the business impact metrics you report to clients in your Monthly AI Performance Report. Where previously your team had to aggregate monitoring data from multiple platform-specific dashboards, Agent 365 provides a unified data stream that can be exported to Power BI, Azure Monitor, or your monitoring platform of choice. Report generation time drops significantly when your monitoring infrastructure has a single authoritative data source.

Function 5: Security and Compliance Controls. Agent 365 integrates with Microsoft Purview, Microsoft Entra ID, and Microsoft Defender to enforce security and compliance controls at the agent level. Sensitivity labels govern what data classifications agents can access. Conditional access policies control which users can interact with which agents. Data loss prevention policies prevent agents from transmitting sensitive data to unauthorized destinations. For healthcare and financial services clients operating under HIPAA, SOC 2, or FINRA requirements, these integrated compliance controls are frequently the most compelling feature of the platform in client conversations.

Function 6: Lifecycle Management. Agent 365 manages the full lifecycle of every agent in the tenant: creation approval, deployment authorization, version tracking, performance monitoring, decommissioning, and audit trail preservation. When a platform vendor updates the underlying AI model, Agent 365 flags the affected agents and prompts a behavioral review before the update propagates. When an agent is retired, Agent 365 preserves the audit trail and decommissions the connected permissions cleanly. This lifecycle discipline is what prevents agent estates from accumulating shadow deployments, expired integrations, and undocumented automations that create security exposure and compliance gaps.

How Agent 365 Fits Into Your Managed Intelligence Stack

Agent 365 slots into your delivery model at two specific points: as the primary governance and monitoring platform within your Monitor and Govern pillars, and as the central control surface through which your Agent Operations Manager manages the client's entire agent estate.

Your technology stack for a standard Managed Intelligence engagement, with Agent 365 integrated, looks like this: Copilot Studio or OpenAI Agents SDK as the agent build environment; Agent 365 as the management and governance control plane; Azure Monitor or Datadog as the supplementary monitoring layer for non-Microsoft agent components; and Power BI as the reporting layer that pulls telemetry from Agent 365 to populate your Monthly AI Performance Reports.

This stack is not theoretical. It is the architecture that enterprise AI teams are standardizing on for Microsoft-adjacent environments, and it is the architecture that gives your MSP practice the operational maturity to win deals against competitors who are managing agents ad hoc through individual platform interfaces.

Positioning Agent 365 in Client Conversations

Most of your clients have never heard of Agent 365. That is an advantage, not a problem. When you introduce it in a sales or advisory conversation, you are demonstrating that your practice is operating at a level of technical sophistication that no competitor has shown them.

The positioning is straightforward. When a client asks how you ensure that their AI agents are being managed professionally, your answer is specific: you use Microsoft's own enterprise agent management platform to maintain continuous visibility across their entire agent environment, enforce governance controls at the policy level, generate the monitoring data that feeds their monthly performance report, and manage the full lifecycle of every deployed agent. That is not a vendor pitch. That is a capability description that most enterprise IT teams would recognize as best-practice agent operations.

For regulated-industry clients, lead with the compliance controls. The integration of Agent 365 with Microsoft Purview sensitivity labels and Microsoft Entra conditional access means that every agent interaction is operating within the same compliance framework the client already uses for their Microsoft 365 data governance. You are not adding a separate compliance layer. You are extending existing compliance infrastructure to cover their agent estate.

Agent 365 vs. Copilot Studio: Knowing the Difference Closes Deals

MSPs who understand the distinction between Agent 365 and Copilot Studio communicate more clearly in client conversations and scope engagements more accurately. The table below captures the essential distinction.

Dimension	Copilot Studio	Agent 365
Primary Role	Build and configure AI agents	Manage, govern, and monitor deployed agents
Where It Lives	Power Platform / M365 admin center	M365 admin center, Microsoft Intune portal
User	Agent builders and developers	IT admins and MSP operations teams
Scope	Individual agent configuration	Entire tenant agent estate
Key Output	A functioning deployed agent	Governance posture, observability data, lifecycle records
Compliance Role	Build-time policy configuration	Runtime policy enforcement and audit trail
Reporting	Individual agent activity logs	Unified telemetry across all agents
MSP Value	Delivery tool for Deploy pillar	Core platform for Monitor and Govern pillars

Getting Started with Agent 365 in Your Practice

Access to Agent 365 management capabilities is available to Microsoft 365 E3 and E5 tenants through the Microsoft 365 admin center under Settings > AI and Copilot. As an MSP, you access your clients' Agent 365 environment through delegated admin access, the same mechanism you use to manage their other Microsoft 365 services.

Your first step is to enable Agent 365 inventory in each client tenant and run the initial agent discovery scan. In most Microsoft-heavy environments, this scan will surface agents end users have created through Microsoft Copilot that no one in IT is aware of. That shadow agent discovery conversation is frequently your most compelling entry point for the Managed Intelligence service: you show the client what is already running in their environment, unmonitored and ungoverned, and you offer to bring it under professional management.

Within 30 days of enabling Agent 365, you should have a complete inventory for each client, baseline telemetry configured for every discovered agent, governance policies drafted for Tier 1 agents, and the Agent 365 data feed connected to your reporting infrastructure. This 30-day activation positions you to deliver your first Monthly AI Performance Report with genuine, platform-sourced data rather than manually aggregated estimates.

Pro Tip

Make the Agent 365 discovery scan your standard onboarding step for every existing managed services client, not just Managed Intelligence prospects. Finding unmanaged agents in a client's Microsoft 365 tenant is a conversation starter that opens the door to the Managed Intelligence pitch without requiring a cold approach. You are showing them a risk they did not know they had and positioning yourself as the solution.

Chapter 2-B Action Items

1. Enable Agent 365 in your own MSP's Microsoft 365 tenant this week. Run the agent discovery scan. The firsthand experience of seeing what Agent 365 surfaces in your own environment is the fastest way to internalize what it will surface in your clients' environments.
2. Enable delegated Agent 365 access in your top three Managed Intelligence prospects' tenants. Run the discovery scan and document every agent found. Use the findings as the opening of your Managed Intelligence assessment conversation.
3. Map your current monitoring infrastructure against the Agent 365 telemetry outputs. Identify how Agent 365 data connects to your reporting platform and what configuration is needed to automate the data flow into your Monthly AI Performance Report template. Complete this configuration before your first production client goes live.

CHAPTER TWO-C

Building Your First Agent: A Hands-On Walkthrough

A Step-by-Step Build of a Helpdesk Ticket Triage Agent in Microsoft Copilot Studio

Every chapter before this one has described what agents do, how they fail, and how to manage them. This chapter shows you how to build one. Specifically, you will walk through the complete construction of a helpdesk ticket triage agent in Microsoft Copilot Studio, from opening the builder for the first time through publishing the finished agent to Microsoft Teams.

This walkthrough is designed for the MSP owner or practice lead who needs to understand what building an agent actually involves, even if a technical team member will execute it. Read through it once so you understand the sequence. Then hand it to your agent engineer as the starting specification for your first internal pilot. By the time you finish this chapter, you will know exactly what questions to ask, what decisions require your input, and what good looks like at each stage.

The agent we are building is a Tier 1 helpdesk ticket triage agent. When a support ticket arrives by email, this agent reads the ticket, classifies it by issue type and urgency, assigns it to the correct queue, drafts an acknowledgment to the end user, and logs the action in your PSA. It handles the first five minutes of every ticket without a technician touching it. In a practice running 200 tickets per week, that is more than 16 hours of technician time recovered, every week.

Build Scenario

Client: Meridian Accounting Group (38 staff, 3 locations). Ticket volume: approximately 45 per week. Current state: all tickets arrive by email, a senior technician manually reads and routes each one, average routing time 4 to 6 minutes per ticket. Target state: agent handles classification and routing autonomously for 70 percent of tickets, escalates the remaining 30 percent with a structured summary for the technician. Platform: Microsoft Copilot Studio with Microsoft 365 E3 licensing already in place.

Before You Open Copilot Studio

Three things must be in place before you touch the builder. Skipping any one of them causes problems mid-build that are harder to fix than preventing.

Access and licensing. The person building the agent needs a Microsoft 365 account with Power Platform permissions enabled. In most MSP environments, this means the account must be assigned the Environment Maker role in the Power Platform admin center. If you are building for a client, you need delegated admin access to their tenant. Confirm both before scheduling the build session.

A defined scope document. Never open the builder without a written scope. At minimum, write down: what triggers the agent (the input), what the agent does with that input (the decision logic), what systems it reads from or writes to (the integrations), and what it is not allowed to do without human approval (the guardrails). One page is enough. The scope document is what you configure the agent against. Without it, you make configuration decisions on the fly and the agent reflects those improvised decisions.

Test data. Before you build, collect 20 to 30 real example tickets from the client. Sanitize them if needed (remove personal data). These become your test cases. You need examples of standard tickets, edge cases, and tickets that should always escalate. Trying to test an agent with no real examples produces false confidence.

> **Pro Tip**
> Do the scope document in a 60-minute session with the client's process owner, the person who currently reads and routes tickets manually. Ask them to walk you through the last ten tickets they handled. The patterns they describe become your agent's decision logic. The exceptions they flag become your escalation triggers.

The Six-Step Build

STEP 1

Estimated time: 15 minutes

Open make.powerapps.com and sign in with the client's tenant credentials (or your delegated admin account). Select the correct environment from the top-right dropdown. This matters: agents built in the wrong environment do not have access to the client's data and cannot be published to the client's Teams channels.

From the left navigation, select Copilot Studio. On the Copilot Studio home screen, click Create. You will see three options: Start from a template, Start from a description, and Start from blank. For your first agent, choose Start from blank. Templates are useful shortcuts once you understand the underlying structure. Starting from blank means you control every configuration decision.

Name the agent precisely. Use the naming convention established in your Agent 365 governance policy: CLIENT-FUNCTION-NUMBER. For this build: MERIDIAN-HELPDESK-001. Add a description that states the agent's purpose in one sentence: "Classifies, routes, and acknowledges inbound helpdesk support tickets." This description appears in Agent 365 inventory and in your Monthly AI Performance Report. Make it accurate.

Leave the language setting as English (United States) unless the client's ticket volume includes other languages. Set the primary channel to Microsoft Teams. Click Create.

✓ Governance Checklist

- ☐ Agent named according to CLIENT-FUNCTION-NUMBER convention
- ☐ Description written and matches scope document
- ☐ Built in correct client tenant environment (not your MSP environment)
- ☐ Correct channel selected (Teams)
- ☐ Agent 365 inventory enabled for this tenant before proceeding

STEP 2

Estimated time: 30 to 45 minutes

The Instructions panel is where you define how the agent thinks and behaves. This is the most important configuration in the entire build. Everything the agent does flows from what you write here. Take your time.

Copilot Studio displays the Instructions panel as a large text field with the placeholder "Describe what your copilot does." What you write here is not a description for humans. It is the operating brief for the AI. Write it the way you would brief a new employee on their first day: specific, structured, and explicit about what they should and should not do.

For the MERIDIAN-HELPDESK-001 agent, your instructions should cover five areas in this order:

1. Role definition. "You are a helpdesk ticket triage agent for Meridian Accounting Group. Your job is to read incoming support requests, classify them by issue type and urgency, assign them to the correct support queue, draft a professional acknowledgment to the submitter, and log the classification in the ticketing system."
2. Classification logic. Define your issue types explicitly. "Classify each ticket into one of these categories: Password and Access (user cannot log in, password reset needed, account locked), Software Error (application not working, error messages, crashes), Hardware Issue (device not functioning, peripheral problems), Network Connectivity (cannot connect to internet or internal resources, VPN issues), New Request (software installation, new equipment, account provisioning), or Other (anything that does not fit the above categories)."
3. Urgency rules. "Assign urgency as follows: Critical (system down affecting multiple users or a time-sensitive client deadline), High (single user completely blocked from working), Medium (user can partially work but is impaired), Low (request or question with no immediate work impact). Default to Medium if urgency cannot be determined from the ticket content."
4. Queue routing. "Route to Tier 1 queue: Password and Access (Low or Medium), Network Connectivity (Low). Route to Tier 2 queue: Software Error, Hardware Issue, Network Connectivity (High or Critical), Other. Route to Account Manager queue: New Request. Escalate immediately to on-call technician: any Critical ticket."
5. Escalation triggers. "Do not process autonomously if any of the following conditions are present: the ticket mentions a data breach, ransomware, or security incident; the submitter is a C-suite executive; the ticket references a regulatory deadline or audit; the ticket is written in a language other than English. Flag these for immediate human review with the reason for escalation clearly stated."

After writing the instructions, click the Test panel on the right side of the screen. Paste in three of your sample tickets and observe the agent's responses. You are looking for accurate classification, correct queue routing, and appropriate escalation behavior. If the classification is wrong, revise the instructions before continuing. Most instruction problems are ambiguity problems: the agent did not know which rule to apply because the rules overlapped.

Watch Out

Do not use vague language in instructions. "Handle tickets appropriately" means nothing to an AI agent. Every decision the agent makes must be traceable to an explicit rule in the instructions. If a behavior is not explicitly defined, the agent will invent a plausible default. That default may not match what you or the client expects.

✓ Governance Checklist

- ☐ Instructions written in five sections: role, classification, urgency, routing, escalation
- ☐ Every issue type has an explicit queue assignment
- ☐ Escalation triggers listed with specific conditions, not general guidance
- ☐ Instructions tested against at least 5 sample tickets before proceeding
- ☐ No vague or open-ended directives in the instruction text

STEP 3

Estimated time: 2 to 4 hours

Integration is where most first-time agent builds slow down. This step takes longer than steps 1, 2, and 4 combined. Plan accordingly. The integration layer is what allows the agent to read from and write to real business systems. Without it, the agent operates only within Copilot Studio and produces outputs that no one in the client's workflow ever sees.

For MERIDIAN-HELPDESK-001, you need three integrations:

Integration 1: Email inbox (ticket source). The agent needs to read inbound emails from the client's support email address. In Copilot Studio, navigate to Actions in the left panel, then Add an action, then select the Microsoft 365 Outlook connector. Authenticate with the service account that monitors the client's support inbox (e.g., support@meridianaccounting.com). Configure the trigger as "When a new email arrives in a shared mailbox." Set the folder to Inbox and enable "Include attachments" if your tickets often come with screenshots or logs.

Test the email connection before moving to the next integration. Send a test email to the support address and confirm the agent receives it and parses the subject and body correctly. If the connection fails, the most common causes are OAuth token configuration and shared mailbox permission settings. Resolve these completely before proceeding.

Integration 2: PSA for ticket logging. The agent needs to create and update tickets in the client's PSA. The specific connector depends on the PSA: ConnectWise Manage, Autotask, HaloPSA, and Syncro all have Power Platform connectors available in the connector library. Search for your PSA by name in the Actions panel. Authenticate with a service account that has

Create and Update permissions on tickets. Configure the action to create a new ticket with the classification, urgency, and queue assignment the agent determined in Step 2. Map the agent's output fields to the PSA's ticket fields: subject, description, category, priority, and assigned queue.

If your PSA does not have a native Power Platform connector, use the HTTP action in Power Automate to call the PSA's REST API directly. This requires your agent engineer to know the PSA's API documentation and authentication method. Budget an additional 2 to 4 hours for this path.

Integration 3: Teams notification for escalations. When the agent identifies a Critical ticket or an escalation trigger condition, it needs to notify the on-call technician immediately. In the Actions panel, add a Microsoft Teams connector. Configure it to post a structured message to your on-call Teams channel. The message should include the ticket subject, the submitter's name and contact information, the escalation reason in plain language, and a direct link to the ticket in the PSA.

After all three integrations are configured, run an end-to-end test with a live email. Send a sample ticket to the client support inbox and trace the output: the agent receives it, classifies and routes it, logs a ticket in the PSA, sends an acknowledgment to the submitter, and (if it meets escalation criteria) posts to the Teams channel. Every step of this chain must work before you proceed.

✓ Governance Checklist

- ☐ Email connector authenticated with dedicated service account, not personal credentials
- ☐ PSA connector tested: ticket creation and field mapping confirmed
- ☐ Teams escalation channel confirmed and on-call notification tested
- ☐ All service account credentials documented in the Operations Runbook
- ☐ Integration health check added to daily monitoring dashboard in Agent 365
- ☐ OAuth token expiry schedule documented (most tokens expire every 60 to 90 days)

Pro Tip

Use a dedicated service account for every agent integration, never a personal user account. When a personal account is disabled, changed, or leaves the organization, every agent connected to it loses its integrations simultaneously. Service accounts are independent of individual users, easier to manage permissions for, and auditable. Set a calendar reminder for 60 days before each service account password expiry.

STEP 4

Estimated time: 45 minutes

Guardrails are the technical controls that enforce the governance policy you wrote before the build. They are separate from the instructions, which tell the agent what to do. Guardrails define the outer limits of what the agent is permitted to do at all.

In Copilot Studio, navigate to Settings then Security. Configure the following:

1. Content moderation. Enable the built-in content moderation filter. This prevents the agent from generating or passing through inappropriate content in its acknowledgment emails. Set the threshold to Medium for a business environment.
2. Authentication. Under Authentication, set the agent to require authentication for any user interaction. This means only authenticated Microsoft 365 users in the client's tenant can trigger the agent or receive its outputs. The agent cannot be accessed by anonymous external parties.
3. Topic fallback. Navigate to Topics and find the default Fallback topic. This topic fires when the agent encounters an input it does not know how to handle. Edit the fallback response to route the ticket to Tier 2 human review with the message: "I was unable to classify this request automatically. A technician will review and respond within 4 business hours." Never leave the default fallback in place, which typically says "I'm sorry, I didn't understand that." A client receiving that message in response to a support ticket loses confidence immediately.
4. Response templates. Write the acknowledgment email the agent sends to every ticket submitter. Keep it professional and specific. Include the ticket reference number (pulled from the PSA after ticket creation), the assigned urgency level, and the expected response time based on urgency. Do not use generic language like "We have received your request." That tells the client nothing. Tell them what you know: "Your request has been classified as a Medium priority Software Error and assigned to our Tier 2 support queue. You will receive an update within 4 business hours."

✓ Governance Checklist

- ☐ Content moderation enabled at Medium threshold
- ☐ Authentication set to require Microsoft 365 tenant login
- ☐ Fallback topic customized: no generic error messages to clients
- ☐ Acknowledgment template includes ticket number, urgency, and response SLA
- ☐ All guardrail settings documented in Governance Policy document for this agent
- ☐ Guardrail configuration screenshot saved to client file for audit record

STEP 5

Estimated time: 2 to 3 hours

Do not skip this step and do not abbreviate it. Testing is where you discover the gap between what you intended and what the agent actually does. That gap always exists on first build. Your job in this step is to close it before the client sees the agent.

Run every one of your 20 to 30 sample tickets through the agent and record the results in a test log. For each ticket, document: the expected classification, the actual classification the agent produced, whether routing was correct, whether the acknowledgment email was appropriate, and any escalation behavior. A simple spreadsheet works. What matters is that you have a written record of every test case outcome.

Focus especially on your edge cases and escalation triggers. These are where agents fail most often. If a ticket about a possible phishing email does not trigger the security escalation rule, fix the instruction before proceeding. If a ticket from the managing partner gets routed to the standard Tier 1 queue instead of escalating, add the condition that handles executive submitters explicitly.

After completing the full test run, calculate your autonomous accuracy rate: the percentage of tickets the agent classified and routed correctly without human correction. A well-configured first-build agent should reach 85 percent or better on standard tickets. If you are below 80 percent, return to Step 2 and refine the instructions. Do not proceed to UAT below 80 percent.

What Good Testing Looks Like

A technician at one MSP ran 28 test tickets through their first triage agent and found that tickets containing the word "slow" were being classified as Hardware Issue instead of Software Error because the instructions did not distinguish between slow device performance (hardware) and slow application performance (software). Two sentences added to the classification logic fixed it. The issue would have generated misfiled tickets for months if not caught in testing. This is exactly what your test library is for.

Schedule a User Acceptance Testing (UAT) session with the client's process owner, the person who currently routes tickets manually. Walk them through 10 tickets in the test environment. Ask them to flag anything that does not match how they would have handled it. Their corrections are not criticism. They are subject matter expertise that improves the agent's accuracy. Incorporate every correction before going live.

✓ Governance Checklist

- ☐ Full test library of 20+ tickets run and results documented
- ☐ Autonomous accuracy rate calculated and above 80 percent before UAT
- ☐ All edge cases and escalation triggers tested with specific examples
- ☐ UAT session completed with client process owner
- ☐ All UAT corrections incorporated and retested
- ☐ Test log retained in client file as part of deployment documentation

STEP 6

Estimated time: 30 minutes plus monitoring

Publishing is the final configuration step. In Copilot Studio, navigate to Publish in the left panel. Click Publish. Copilot Studio validates the agent configuration and, if all required fields are complete, publishes the agent to the environments you have configured. You will see a green confirmation when publishing is complete.

To make the agent available in the client's Microsoft Teams environment, navigate to Channels in the left panel and select Microsoft Teams. Click Open in Teams. This opens the Teams admin center where you configure which Teams channels and users the agent is available to. For the pilot phase, restrict availability to the support team only. Full organizational rollout follows after 30 days of confirmed pilot performance.

After publishing, pin the agent to your monitoring dashboard in Agent 365. The agent should appear in your tenant inventory within 15 minutes of publication. Confirm that Agent 365 is receiving telemetry from the agent before considering the deployment complete. If telemetry is not flowing within 30 minutes, check the Agent 365 diagnostic settings for the tenant.

Run one live end-to-end test immediately after publishing. Send a real support email to the client's support inbox. Confirm the ticket appears in the PSA, the acknowledgment goes to the submitter, and Agent 365 logs the action. When all three confirm, the agent is live.

✓ Governance Checklist

- ☐ Agent published successfully in Copilot Studio (green confirmation recorded)
- ☐ Teams channel configured and restricted to pilot group for first 30 days
- ☐ Agent visible in Agent 365 tenant inventory
- ☐ Agent 365 telemetry confirmed flowing within 30 minutes of publication
- ☐ Live end-to-end test completed with confirmed PSA ticket, acknowledgment email, and Agent 365 log
- ☐ Operations Runbook updated with go-live date and initial SLA benchmarks
- ☐ Client informed of go-live with expected response times and escalation contact

The First 30 Days After Go-Live

Going live is not the end of the deployment. It is the beginning of the Monitor phase. For the first 30 days, your operations team reviews every agent action daily. You are looking for classification errors, routing failures, acknowledgment emails that went to the wrong person, and any tickets where the escalation trigger should have fired but did not.

Set a performance benchmark at day 7. Compare the agent's autonomous completion rate to your UAT baseline. It should be within 5 percent. If it drops significantly below that, something in the production environment differs from your test environment. Common causes: ticket format differs from test samples (clients often format production tickets differently than how they described them to you), integration latency under live load, or a PSA field mapping that worked in testing but breaks with real ticket data.

At day 30, calculate three metrics: autonomous completion rate (percentage of tickets handled without technician intervention), average classification accuracy (confirmed correct vs. total), and technician time saved per week. These three numbers become the opening data point in your first Monthly AI Performance Report for this agent. They also become the baseline against which every subsequent month is measured.

If your day-30 autonomous completion rate is at or above the 70 percent target defined in the scope document, the pilot is successful. Schedule the 90-Day Business Review with the client to present the results and introduce the Expansion Opportunity List for phase two.

> **Pro Tip**
>
> Tell the client's team what to do when the agent gets something wrong. If a technician sees a misclassified ticket, they should correct it in the PSA and flag it in a shared log that your operations team reviews weekly. Every correction is a data point that drives instruction refinement. Clients who understand they are actively contributing to the agent's improvement become stakeholders in its success rather than critics of its limitations.

What You Can Do With This Walkthrough Right Now

You have just traced the complete lifecycle of an agent build from blank canvas to live production system. Before you read the next chapter, take two actions.

First, identify a process in your own MSP operations that this same agent pattern could address. Helpdesk ticket triage is the example, but the pattern, read input, classify it, route it, acknowledge it, log it, is applicable to contract renewal alerts, new client onboarding requests, vendor invoice processing, and a dozen other high-frequency workflows. Pick one and write the scope document for it. You now have enough knowledge to write a credible first draft.

Second, if you have a technical team member who will own agent builds, walk them through this chapter together. Go step by step and let them raise questions about your specific PSA, your specific Microsoft 365 configuration, and your specific client environments. The questions they ask in that conversation are exactly the questions they would otherwise encounter as blockers mid-build. Answer them now, before the clock is running.

> **Chapter 2-C Action Items**
>
> 1. Write the scope document for your first internal agent build this week. One page: trigger, decision logic, integrations, guardrails. Use the MERIDIAN-HELPDESK-001 scope as your template.
> 2. Enable Copilot Studio in your MSP's own Microsoft 365 tenant if you have not already. Build the helpdesk triage agent for your own support inbox using this walkthrough. Running it on your own tickets before a client's gives your team real experience at zero client risk.
> 3. Collect 20 real support tickets from your target client and classify them manually before building. The patterns you find in that manual exercise define the agent's classification logic. This single step improves first-build accuracy by 20 to 30 percent.

CHAPTER THREE

The Managed Intelligence Framework

Five Pillars, One Practice, Infinite Compounding Revenue

The Managed Intelligence Framework is the operating model of the service. It defines the five practice pillars that every engagement is built on, from the initial opportunity assessment through the ongoing delivery of a continuously improving agent operation. Every client engagement runs through all five pillars. Every pillar generates revenue. Together, they constitute a managed service that creates value for the client and compounding recurring revenue for your practice.

The framework is not a one-time project sequence. It is a continuous operating model. You do not complete Discover and move on. You do not deploy an agent and hand it to the client. You operate the full framework every month, for every client, in perpetuity. That is what makes Managed Intelligence a managed service rather than a consulting engagement.

Pillar	What It Delivers
DISCOVER	Assess the client's AI agent opportunity. Identify the processes worth automating. Build the business case and the deployment blueprint. This is a paid, fixed-fee engagement that generates immediate project revenue and creates the architecture for everything that follows.
DEPLOY	Configure agents, establish system integrations, test rigorously, and go live. Integration-first design means you build reliable connections before you build agent behavior. Phased go-live means you prove the agents work before full production rollout.
MONITOR	Provide real-time operational visibility across every deployed agent. Health, uptime, performance, and business impact tracked continuously. Monthly AI Performance Reports delivered to every client. No unpleasant surprises.
GOVERN	Enforce compliance controls. Maintain audit trails. Manage agent lifecycle. Define and enforce policy guardrails. This pillar is your primary competitive differentiator and the primary reason regulated-industry clients choose you over any competitor.
OPTIMIZE	Improve agent performance continuously. Identify and develop new use cases every 90 days. Grow the client's agent count. Compound your MRR without requiring new client acquisition.

The Operating Cadence

Each pillar runs on a defined cadence. Understanding this cadence is essential for staffing your practice, setting client expectations, and building the operational infrastructure that allows you to scale.

Continuous	Daily	Monthly	Quarterly
Agent health monitoring, integration connectivity, SLA tracking, automated alerting			
Performance log review, exception analysis, governance policy check, internal ops review			
AI Performance Report delivery, optimization recommendations, billing reconciliation, governance audit			
90-Day Business Review, expansion opportunity presentation, roadmap update, client success assessment			

What Managed Intelligence Delivers for Your Clients -- in Plain Language

Your clients pay for Managed Intelligence because it gives them three things they cannot get any other way.

First, they get AI-powered operations without the burden of managing AI. Their team uses the outputs. Your team handles everything else.

Second, they get professional governance and compliance controls. Every agent action is logged, auditable, and operating within defined policy guardrails. In regulated industries, this is not optional.

Third, they get a technology partner who is actively and continuously improving their AI capability -- not just the firm that set it up 18 months ago and sends an invoice every month.

Chapter 3 Action Items

1. Draw your Managed Intelligence service map. On a single page, map the five pillars, the activities within each, and the client-facing deliverables each pillar produces. This document becomes your internal service definition and your external service overview.
2. Define your monitoring platform. Evaluate Microsoft Azure Monitor, Datadog, and agent-platform-native analytics tools. Choose one and standardize on it before your first production deployment.
3. Write your service description in 150 words. Describe what Managed Intelligence is and what it delivers in language your clients' CEOs would understand immediately. This language appears in every proposal and every sales meeting.

PART TWO

The Five-Pillar Delivery Model

A Complete Operational Methodology for Each Pillar

CHAPTER FOUR

Discover

Find the Opportunities Worth Building

Every Managed Intelligence engagement starts with Discover. Before you build an agent, you assess. You understand the client's operational landscape. You identify which processes are worth automating and which are not. You model the ROI. You design the deployment architecture. You define the governance requirements.

Discover is a paid, fixed-fee engagement. Price it at $5,000 to $15,000 depending on client size. Never give it away as a presales activity. A paid assessment signals that you are an expert whose time has value, produces a deliverable the client will use for the next 18 months, and creates the commitment from both parties that makes a successful engagement possible.

The Five Assessment Dimensions

Dimension 1: Process Inventory and Priority Scoring

Identify every significant, recurring process in the client's operation that could benefit from AI agent involvement. Score each process on four dimensions: frequency (how often it occurs), manual effort (how many person-hours it consumes per week), error rate (how often it produces incorrect or inconsistent outputs), and agent suitability (how rule-based the decision logic is and how available the underlying data is).

Processes with the highest combined scores are your Tier 1 agent targets. A client billing process that happens 200 times per week, consumes 30 hours of staff time, produces errors 15% of the time, and follows clear decision rules is a Tier 1 target. A strategy review process that happens quarterly and depends on judgment that changes every cycle is not.

Dimension 2: Data and System Readiness

AI agents need data. The right data. In the right format. From the right systems. With the right permissions. Before you scope any agent deployment, assess which systems the target processes rely on, what API or integration capability those systems have, what data quality issues exist, and what access credentials and permissions will need to be configured.

Data readiness gaps identified during assessment become Horizon 1 preparatory work. Better to find them in a structured assessment than to discover them mid-deployment when the timeline and budget are already committed.

Dimension 3: Governance and Compliance Requirements

Before you deploy anything, you need to know the regulatory environment the agent will operate in. What compliance frameworks apply? What data will the agent access and process? What are the client's internal approval requirements for automated decision-making? These answers determine the governance architecture you design before a single line of agent configuration is written.

Dimension 4: Workforce Readiness

Agents change how people work. Some of your client's staff will welcome the change. Others will resist it. The Discover assessment includes a structured evaluation of which teams are affected, what their current comfort level with AI tools is, where resistance is likely, and what change management investment is needed to drive adoption.

A technically perfect agent that the team does not trust or use is a failed deployment. You cannot fix adoption problems after go-live. You need to design for them before.

Dimension 5: ROI Modeling

The assessment concludes with a quantified ROI model for each Tier 1 opportunity. The model starts with the current cost of the process -- person-hours per week multiplied by the fully loaded labor rate. It projects the impact of agent deployment -- typically a 50% to 75% reduction in human time for Tier 1 candidates based on industry benchmarks. It calculates the implementation investment and ongoing management fee. It produces a breakeven timeline and a net present value over 24 months.

When a CFO sees that a three-agent deployment costing $28,000 to implement and $3,200 per month to manage is saving their team $7,500 per month in labor cost, the conversation shifts from 'Is this worth it?' to 'When can we start?'

The Discover Deliverables

Agent Opportunity Report: A written assessment covering the process inventory and priority scores, system and data readiness findings, governance and compliance requirements, workforce readiness evaluation, and ROI model for each Tier 1 opportunity.

Managed Intelligence Deployment Blueprint: A phased deployment plan specifying which agents will be built in what sequence, the platform for each, the system integrations required, the governance policies that will govern each agent, and the performance benchmarks at 30, 60, and 90 days post-deployment.

These two documents are what the client is paying for. They are also what you will reference for every decision made in the Deploy, Monitor, Govern, and Optimize phases. Done properly, the Discover deliverables make every subsequent phase faster, cheaper, and less risky.

Chapter 4 Action Items

1. Build your Agent Opportunity Scoring Template. Create a structured worksheet that scores each candidate process on the four dimensions using a 1 to 5 scale. Total scores drive tier assignment: 16 to 20 is Tier 1, 10 to 15 is Tier 2, below 10 is deprioritized.

2. Build your ROI model template. A simple spreadsheet that calculates current process cost, projected agent savings, implementation investment, and breakeven timeline. This model appears in every Discover deliverable.

3. Write your Discover engagement proposal. Use the assessment deliverables defined above as your scope of work. Price it at $5,000 for clients under 25 users and scale from there. Practice presenting it out loud before your first sales meeting.

CHAPTER FIVE

Deploy

Go Live Without the Drama

Deploy is where the assessment becomes an operating agent. It is also where most MSPs who try to build AI practices without a structured methodology run into trouble. They rush the integration work. They skip UAT. They go live before the governance framework is ready. Then they spend the next three months firefighting instead of optimizing.

The discipline you apply during Deploy directly determines your client satisfaction, your operational cost, and your capacity to scale. Do it right once and you have a repeatable model. Rush it and you have a recurring operational problem.

The Integration-First Principle

Every Managed Intelligence deployment follows one governing principle: integration first. Every time. No exceptions.

The integration infrastructure is the foundation of agent performance. A sophisticated agent with a fragile or unreliable connection to its data sources will fail in production even if its reasoning logic is excellent. Build the integrations first. Test them independently of the agent logic. Confirm they are stable before you write a single line of agent configuration.

In practice, this means Phase 1 of every deployment is exclusively integration work. You establish API connections, configure authentication, set data access permissions, validate data flows in both directions, and confirm that every system the agent will touch is accessible, stable, and returning data in the expected format. No agent configuration happens until the integration layer is fully tested and passing.

The Four-Phase Deployment

Phase 1: Integration Architecture (Weeks 1 to 2)

Map every system the agent will interact with. Establish API connections. Configure authentication and authorization. Set data access permissions. Test data flows. Identify and resolve data quality issues. Establish the monitoring infrastructure that will track agent health from day one.

Deliverable: A tested, stable integration architecture. Nothing moves to Phase 2 until this is complete.

Phase 2: Agent Configuration and Training (Weeks 2 to 4)

Configure the agent's core behaviors: its instructions, decision rules, escalation protocols, error handling, and communication templates. Work closely with the client's process owners – the people who actually perform the process today – to validate that the agent's logic matches real-world operational reality. Their input is not optional. They are the subject matter experts. Treat them that way.

Configure the governance guardrails during this phase, not after go-live. Define what the agent is permitted to do autonomously, what requires human review, and what triggers an escalation. Document everything in the Governance Policy document for this agent.

Phase 3: Testing and User Acceptance (Weeks 3 to 5)

Test against a comprehensive set of cases: standard scenarios that represent 80% of normal volume, edge cases that represent the unusual but valid situations the agent will encounter, and known exceptions that represent the situations where the agent should always escalate to human review.

Conduct a structured UAT session with the client's process team. Walk them through each test scenario. Capture their feedback. Resolve every issue they identify before advancing. The team members who validate the agent in UAT are far more likely to trust it in production. This is not just quality control. It is change management.

Phase 4: Controlled Go-Live (Weeks 5 to 6)

Launch with a defined pilot group before full organizational rollout. Monitor every agent action during the pilot with daily review. Address behavioral issues immediately. Confirm that monitoring alerts are correctly configured and that the operations team can respond effectively.

Full rollout follows the pilot only when pilot metrics confirm performance meets the benchmarks in the Deployment Blueprint. Adding one to two weeks for a controlled go-live is always worth it. A high-visibility failure in full production is always worse than a brief delay.

The Most Common Deployment Failures

Failure Mode	Root Cause and Prevention
Integration brittleness	API connections that work in testing fail in production. Rate limits, token expiry, and schema changes are the usual culprits. Build explicit error handling and automated retry logic into every integration. Monitor connectivity as a primary health metric.
Scope additions mid-build	Clients see what is being built and want to add requirements that were not in the original blueprint. Enforce a formal change control process. Any new requirement requires scope assessment and additional pricing before implementation begins.
Insufficient test coverage	Agents fail on edge cases that were not tested. Build your test library systematically: 60% standard scenarios, 25% edge cases, 15% known exceptions. Do not advance to UAT until all three categories are covered.
Going live under schedule pressure	Business urgency pushes the team to launch before validation is complete. Set go and no-go criteria in writing before deployment begins. Launch when the criteria are met, not when someone is impatient.

Chapter 5 Action Items

1. Build your Deployment Checklist. A master go-live checklist covering integration testing, agent behavior validation, UAT sign-off, monitoring activation, governance policy completion, and client training. Nothing goes live without every item checked.

2. Develop your test case library. For your first three agent types, build 20 or more test cases each covering standard, edge, and exception scenarios. This library improves with every engagement.

3. Write your Change Control Policy. Define in one page how scope changes are requested, assessed, priced, and approved during deployment. Include this policy in every engagement agreement.

CHAPTER SIX

Monitor

See Everything Before Your Client Does

Monitoring is what makes Managed Intelligence a managed service. It is the continuous operational presence that justifies your monthly fee, generates the data that proves your value, and catches every issue before it becomes a client problem. Without rigorous monitoring, you are not managing agents. You are hoping they keep working.

Unmonitored agents fail quietly. They do not crash with an obvious error. They drift. Their performance degrades gradually. Their integrations develop small cracks that grow into full failures over weeks. By the time the client notices something is wrong, the problem has been building for months. At that point, you have no performance history, no early warning data, and no evidence that active management was happening. That situation is a client relationship crisis and a reputation problem.

Active monitoring prevents all of it.

Three Monitoring Layers

Layer 1: Agent Health Monitoring

At the most basic level, you are monitoring whether each agent is operational. Health metrics cover uptime percentage (target: 99.5% or better for production agents), integration connectivity status (are all connected systems reachable and responding?), processing latency (is the agent completing tasks within the expected time window?), and error rate (what percentage of inputs are generating processing errors?).

Health monitoring runs continuously and automatically. Your platform checks every agent every five minutes. When any health metric falls below its defined threshold, an alert fires to your operations team. No human attention is needed until an alert fires.

Layer 2: Performance Monitoring

Beyond basic health, you track whether each agent is doing its job well. Performance metrics vary by agent type but consistently include task completion rate (what percentage of inputs does the agent handle autonomously versus escalating to human review?), accuracy rate (for agents whose outputs can be validated, what percentage are correct?), throughput (how many tasks per day or week is the agent processing?), and SLA compliance (are defined response time commitments being met?).

Performance monitoring requires human review -- not of every agent action, but of aggregate performance data on a daily basis. Your operations team checks dashboards daily and investigates any metric that has moved outside of normal bounds by more than 10%.

Layer 3: Business Impact Monitoring

This is the layer that makes your client reports compelling. Business impact metrics translate agent performance into business language: hours saved per week, errors eliminated per month, cost savings in dollars, and cycle time improvement as a percentage reduction. These are the numbers that appear in your Monthly AI Performance Report. These are the numbers a client CEO shows to their board. These are the numbers that make the management fee look like an obvious investment when the contract comes up for renewal.

The Monthly AI Performance Report

Every Managed Intelligence client receives a Monthly AI Performance Report by the fifth business day of the following month, every month without exception. This report is the most visible deliverable in your service and the primary tool through which you demonstrate ongoing value.

The report structure covers five sections: an executive summary (one page covering overall health, top highlights, any significant incidents, and the month's primary business impact metric), agent performance scorecards (one page per agent with all three monitoring layers presented visually), an incident log (every alert, exception, and intervention during the reporting period), optimization recommendations (two or three specific, evidence-based recommendations for the next period), and roadmap status (progress on active development projects and next milestone dates).

Pro Tip

Automate the data collection and initial population of your Performance Report template. Power BI or your monitoring platform's export capability can pull the quantitative sections automatically. Your team's time should go into interpretation and recommendations -- the parts that require human judgment -- not into copying numbers from dashboards into a document.

Chapter 6 Action Items

1. Select your monitoring platform. Evaluate Microsoft Azure Monitor, Datadog, and at least one agent-platform-native option. Make a decision and standardize before your first production client. Switching platforms mid-delivery is expensive.

2. Build your Monthly AI Performance Report template. Design the format -- sections, visual presentation, key metrics -- before you have clients to report on. Having the template ready makes your first report faster and ensures consistency across clients.

3. Define alert thresholds in writing. For every health and performance metric, specify the exact value that triggers an alert. Document these in your operations runbook so every team member applies them consistently.

CHAPTER SEVEN

Govern

Build the Moat Your Competitors Cannot Cross

Any technically capable team can deploy an AI agent. Building and maintaining the governance framework that makes agent deployment safe, compliant, auditable, and defensible in a regulated environment is a different capability entirely. And that capability is what most of your competitors do not have.

Governance is your primary competitive differentiator in Managed Intelligence. It is the capability that wins regulated-industry clients. It is the reason your service cannot be easily displaced by a vendor going direct or a competitor copying your surface-level offering. Governance creates depth that takes years to build, and that depth compounds with every engagement you complete.

Five Governance Elements

Element 1: Policy Guardrails

Every deployed agent operates within a documented policy framework that specifies what the agent is permitted to do autonomously, what it is explicitly prohibited from doing, and what conditions require human review before action is taken. These guardrails are not just written policies. They are technical controls configured at the platform level that the agent cannot override.

Key guardrail categories: data access policies (which data sources the agent can access and under what conditions), action authorization limits (what actions the agent can take without human approval), escalation triggers (conditions that cause the agent to pause and route to a human reviewer), communication authorization (what messages the agent can send on behalf of the client and what requires human approval), and scope boundaries (the explicit definition of the business processes within this agent's operational scope).

Element 2: Audit Trails

Every action taken by every deployed agent must be logged, timestamped, and retained in a format that supports compliance reporting and incident investigation. The audit log captures: the input that triggered the action, the decision path the agent applied, the action taken, the output produced, and the outcome including any human review or override.

For clients subject to HIPAA, SOC 2, GDPR, or other regulatory frameworks, audit logs must meet specific retention requirements and access controls. Your ability to say 'we maintain a complete, immutable log of every action your agents take, accessible for compliance review at any time' is frequently the statement that closes deals in healthcare, financial services, and legal services.

Element 3: Lifecycle Management

AI agents are not static. The AI models they run on are updated by platform vendors. The business systems they integrate with release new versions. The business processes they serve evolve. Without formal lifecycle management, agents that were accurate and appropriate at deployment drift into inaccuracy and inappropriate behavior over time.

Lifecycle management covers: model version reviews (when the platform vendor updates the underlying AI model, how does agent behavior change?), integration compatibility monitoring (do connected system updates break the integrations?), process change protocols (when the business process changes, how is the agent updated?), and periodic behavioral audits (independent review of agent outputs to confirm they remain aligned with intended behavior and policy guardrails).

Element 4: Incident Response

When an agent behaves unexpectedly -- taking an action outside its intended scope, producing systematically incorrect outputs, experiencing a significant performance degradation -- you need a documented, practiced incident response process. This process must be in your operations runbook, reviewed by your team quarterly, and communicated to clients at the start of every engagement.

Your Incident Response Protocol covers: detection (how the incident is identified), triage (initial severity and scope assessment), containment (isolating the affected agent or pausing operation), investigation (root cause analysis), remediation (fixing the underlying issue), and post-incident review (governance policy updates to prevent recurrence).

Element 5: Regulatory Compliance

AI regulation is advancing at state, federal, and international levels. The EU AI Act, US federal AI executive orders, state privacy laws, and sector-specific requirements are creating an overlapping compliance landscape that most SMBs are completely unprepared to navigate. Your Managed Intelligence service includes ongoing regulatory monitoring and compliance advisory as a core component.

Regulatory Framework	Key Agent Compliance Implications
HIPAA	Patient data handling, audit trail requirements, access controls, business associate agreements for AI vendors
SOC 2	Security, availability, processing integrity, and privacy controls for agent operations
GDPR and State Privacy Laws	Data minimization, consent management, right to explanation for automated decisions
PCI DSS	Payment data isolation, access controls, audit logging for agents touching financial transactions
CMMC	Controlled unclassified information handling and audit trails for defense-adjacent clients
EU AI Act	Risk classification for AI systems, transparency requirements, human oversight mandates by risk tier

Watch Out: EU AI Act Classification

The EU AI Act classifies certain AI agent use cases as high-risk systems subject to mandatory conformity assessments, audit requirements, and transparency obligations. If any of your clients operate in EU markets or process data from EU residents, verify whether their agent use cases fall within the high-risk classification before deployment begins.

Chapter 7 Action Items

1. Build your Policy Guardrail Template. Create a standard policy document covering the five guardrail categories for a typical agent deployment. Customize it for each client and each agent type before go-live.

2. Select your audit logging platform. Your audit trail must live in a dedicated, immutable logging system -- not spreadsheets or email records. Azure Monitor Logs, Splunk, and Datadog Logs are the three most common choices for MSP environments.

3. Write your Incident Response Runbook. Use Appendix D as your starting point. Walk your operations team through it before your first production client. A practiced response is significantly more effective than an improvised one.

CHAPTER EIGHT

Optimize

Grow Revenue Without Growing Your Client Count

The Optimize pillar is where Managed Intelligence becomes a compounding business. Every month of active agent operation produces data. That data reveals where performance can improve, where new use cases exist, and where expanding the agent scope creates additional value for the client. Every improvement justifies a deeper relationship. Every expansion increases your MRR without requiring a new client acquisition effort.

This is the commercial mechanic that makes Managed Intelligence different from every other MSP service. Your revenue from a Managed Intelligence client does not plateau. It grows continuously as the agent count increases and as performance improvements compound. Managing this dynamic deliberately is one of the most important commercial skills in the practice.

The 90-Day Expansion Cycle

The natural rhythm of Managed Intelligence expansion is 90 days. Agents typically reach their initial performance plateau within 60 to 90 days of deployment. At that point, the optimization work shifts from improving current agents to identifying what to build next. The 90-day cycle also aligns with business planning and budget review cycles, making clients most receptive to expansion conversations at the quarterly mark.

Your 90-Day Business Review is the commercial engine of the Optimize pillar. Every 90 days, you meet with the client's executive sponsor, present the performance summary, demonstrate the business impact created since the last review, and introduce the Expansion Opportunity List -- a prioritized set of the next two or three highest-value agent opportunities identified through your ongoing operational engagement with the client's business.

The key to a productive 90-Day Business Review is that the expansion recommendations are already half-sold before the meeting starts. Over the previous 90 days, your operations team has been paying attention -- noting the processes team members mentioned as painful, the exceptions that appeared repeatedly in the agent logs, the business changes that created new operational challenges. By the time you sit down for the review, you are presenting solutions to problems the client already knows they have.

The Four Optimization Activities

Performance Tuning

Every agent can be improved. As performance data accumulates, patterns emerge. You identify the input types the agent handles well, the scenarios where it consistently underperforms, and the decision rules that generate the most exceptions. Systematic tuning driven by data produces measurable improvement in accuracy rates, autonomous completion rates, and throughput. Every percentage point improvement in autonomous completion translates directly into additional labor savings for the client and additional value evidence for your service.

Prompt and Instruction Optimization

For agents built on large language models, the quality of the agent's instructions has an enormous effect on performance. Prompt optimization is an ongoing discipline: review agent outputs regularly, identify behavioral patterns that are suboptimal, test instruction changes in a controlled environment, and deploy improvements that produce better outcomes. This is a specialized skill your team develops over time, and it is one of the hardest capabilities for clients to replicate on their own.

Integration Maintenance

Every connected system is a potential source of integration drift. API changes, authentication updates, and platform upgrades can degrade agent performance without triggering obvious alerts. Proactive integration maintenance -- scheduled health checks, vendor release note monitoring, and automated integration testing -- catches these issues before they affect agent performance or client operations.

Use Case Development

As deployed agents prove their value and the client's comfort with agent-augmented operations grows, new use cases become viable that were not practical at initial deployment. Your operations team identifies these opportunities continuously and documents them in the client's Expansion Backlog. The Expansion Backlog is a living document, updated every month and reviewed with the client every 90 days.

The Compounding Revenue Math

Here is what the compounding revenue looks like for a single Managed Intelligence client over three years.

Phase	Scope	MRR	ARR
Initial deployment	3 agents: helpdesk triage, invoice processing, client onboarding	$2,400/month	$28,800/year
Year 1 expansion	2 additional agents: compliance reporting, sales qualification	$3,800/month	$45,600/year
Year 2 expansion	3 additional agents plus tier upgrade for orchestration capability	$6,200/month	$74,400/year
Year 3 optimization	Performance improvements, expanded scope, governance retainer	$7,500/month	$90,000/year

One client. No new sales cycle. Revenue growing from $28,800 in year one to $90,000 in year three because the agents are creating real business value and you are consistently identifying new applications of that capability. Multiply this across 10 clients and you have a Managed Intelligence practice generating $700,000 to $900,000 in annual recurring revenue -- built entirely on existing client relationships.

Pro Tip

Maintain a running Expansion Backlog for every Managed Intelligence client. A prioritized list of future agent use cases, updated monthly. When you walk into a 90-Day Business Review with a populated, evidence-backed Expansion Backlog, the expansion conversation is easy. You are presenting solutions to problems the client already knows they have -- not pitching new services.

Chapter 8 Action Items

1. Design your 90-Day Business Review agenda. Create a standard meeting structure with timing: performance summary (20 minutes), business impact demonstration (15 minutes), expansion opportunity presentation (20 minutes), and discussion and next steps (15 minutes). The meeting should always end with a defined next step.

2. Build your Expansion Backlog template. A simple document that captures the opportunity name, the business problem it addresses, the estimated monthly value, the implementation complexity, and the priority tier. Update it monthly for every client.

3. Run the compounding revenue projection for your first three target Managed Intelligence clients. Using the model above, project what those three clients could generate in MRR at the end of year three. Use this number as your practice-building target.

PART THREE

The Commercial Model

Pricing, Selling, Delivering, and Scaling the Practice

CHAPTER NINE

Pricing Managed Intelligence

Charge What It Is Worth

Most MSPs price Managed Intelligence the same way they price traditional managed services -- by the seat, by the device, or by the hour. All three approaches undervalue the service significantly. Managed Intelligence pricing is built on three components: an onboarding investment, a recurring management fee, and an expansion model. Together, these components create a pricing architecture that generates immediate project revenue, establishes high-margin recurring revenue, and rewards expansion with compounding growth.

Component 1: The Onboarding Investment

The onboarding investment is a fixed-fee project covering the Discover engagement and the initial Deploy phase. It is priced on scope -- the number of agents to be deployed, the complexity of the integrations, and the sophistication of the governance framework required.

Onboarding Tier	Investment Range
Starter: 1 to 2 agents, simple integrations (2 to 3 systems)	$5,000 to $12,000
Standard: 3 to 5 agents, moderate integrations (4 to 6 systems)	$12,000 to $28,000
Advanced: 6 to 10 agents, complex integrations	$28,000 to $60,000
Enterprise: 10 or more agents, orchestration tier	$60,000 to $150,000 or more

The recurring management fee covers Monitor, Govern, and Optimize on an ongoing monthly basis. It is tiered by agent count and service depth.

Component 2: The Recurring Management Fee

The recurring management fee covers Monitor, Govern, and Optimize on an ongoing monthly basis. It is tiered by agent count and service depth.

Foundation	Essentials	Professional	Enterprise
1 to 3 agents	**4 to 8 agents**	**9 to 15 agents**	**16 or more agents**
Health monitoring, monthly report, standard governance	All Foundation plus performance monitoring, quarterly optimization, compliance reporting	All Essentials plus real-time alerting, governance retainer, 90-day reviews, dedicated ops manager	All Professional plus custom SLA, advanced governance, regulatory audit support, strategic advisory
$500 to $1,500/month	$1,500 to $4,000/month	$4,000 to $8,500/month	$8,500 to $20,000 or more/month

Component 3: The Expansion Model

Every new agent added to an existing engagement is priced at 60 to 70% of the standard onboarding rate for equivalent complexity (the integration infrastructure is already established) plus an incremental monthly fee at the applicable tier rate. This creates a natural pricing model that rewards clients for growing their agent count while ensuring your revenue grows proportionally.

Value-Based Pricing Validation

Before finalizing pricing for any engagement, validate it against the value being created. Use the ROI model from the Discover phase to calculate the monthly value of the deployed agents. Your management fee should represent 25 to 40% of the monthly value created.

A client whose three agents collectively save $12,000 per month in labor costs is receiving $72,000 in annual value from a $36,000 annual management fee. That is a 2x return that any executive can defend to their board. Price in that range and you will rarely lose a deal on cost.

Commitment Structures

Managed Intelligence is most commercially attractive as a multi-year commitment. Offer a 10% discount for a 24-month agreement and a 15% discount for a 36-month agreement. These discounts are easy to justify -- longer commitments allow you to invest more deeply in the client's agent architecture -- and they improve the lifetime value of each client significantly.

Never offer month-to-month agreements without a 25% premium above the annual rate. Month-to-month removes the commitment that allows you to staff appropriately, invest in the integration architecture, and build the institutional knowledge that makes the service excellent. Price it to reflect that risk.

Chapter 9 Action Items

1. Build your Managed Intelligence pricing calculator. An Excel tool that takes agent count, complexity tier, and commitment term as inputs and outputs the onboarding fee and monthly management fee. Use it in every sales conversation.

2. Define your price floors. Based on your cost structure and margin targets, set the minimum onboarding fee and minimum monthly fee you will accept. Write these down. Discounting below your floor is how practices go out of business.

3. Build your ROI validation template. For every engagement, produce the model showing the client the monthly value of their agent operations and the management fee as a percentage of that value. This model appears in every quarterly report.

CHAPTER TEN

Selling Managed Intelligence

The Conversation That Wins

Selling Managed Intelligence is not the same as selling managed services. The traditional MSP sales approach -- lead with compliance risk, anchor on the cost of downtime, close on a per-seat monthly fee -- does not transfer directly. Clients who buy Managed Intelligence are buying something different: a capability they cannot build themselves, a governance framework they do not have the expertise to maintain, and a competitive position that helps them move faster than their peers.

What the Client Is Actually Buying

When a business owner invests in Managed Intelligence, they are buying three things. Understanding these three things is the foundation of your sales approach.

Operational capacity: The ability to do more work with the same team. AI agents handle the high-volume, repetitive, rule-based work so the client's people can focus on the work that requires judgment, relationships, and expertise. This is immediately compelling to any business owner who feels like growth is constrained by headcount.

Operational confidence: The assurance that their AI agents are operating correctly, compliantly, and with professional oversight. They are not running an AI experiment. They have a managed service with defined SLAs, monthly performance reporting, and an expert team responsible for keeping it running and improving.

Strategic acceleration: The ability to identify and deploy new AI capabilities faster than competitors who are figuring it out alone. Your cross-client experience -- the use cases you have built, the governance approaches you have tested, the performance benchmarks you have established -- is applied to their specific business context without them paying for your learning curve.

The Sales Conversation Framework

Open With the Business Problem

Never open a Managed Intelligence conversation by explaining what AI agents are. Open by asking about the business problem. 'What is the one process in your operation that consumes the most time and creates the most frustration for your team?' That question surfaces the specific operational pain that Managed Intelligence addresses and anchors everything that follows in business value rather than technology features.

Every client you talk to has a process that fits this description. It might be client onboarding. It might be invoice processing. It might be support ticket triage. Your job in the first ten minutes is to find it.

Introduce the Possibility

Once you have identified a specific, painful, high-frequency process, introduce the possibility of agent-based automation without pitching it. 'We have been helping businesses like yours automate exactly this kind of process using AI agents. In a similar situation, we reduced the time your team spends on this by about 70% in the first 60 days. Would it be worth spending 30 minutes exploring whether something similar could work for your business?'

This is not a sales pitch. It is an invitation to a conversation. The threshold to say yes is very low. And a yes here leads directly to the assessment proposal.

Propose the Assessment

The 30-minute exploration conversation ends with a proposal for the Discover engagement. Frame it directly: 'Before we recommend a specific solution, we do the work to understand your specific situation -- your processes, your systems, your compliance requirements. The output is a written assessment that shows you exactly which agents we would build, what value they would create, and what ongoing management looks like. For an organization your size, that assessment is $8,500. Would you like to proceed?'

The transition from exploration to paid assessment is the most important commercial moment in the Managed Intelligence sales process. Everything that follows -- the deployment, the management, the expansion -- flows from the client's commitment to this first engagement.

Handling the Four Most Common Objections

Objection	Response
We are not ready for AI yet.	Most of our clients say exactly this before their assessment. What the assessment typically shows is that they are closer to ready than they think -- and that the gaps are specific and addressable. The assessment tells you what ready looks like for your business specifically. That is the whole point of it.
This seems expensive.	Let me show you the math. The three processes we would target first are collectively consuming approximately X hours per week. At your fully loaded labor rates, that is Y per month in staff time. Our management fee is Z -- which is roughly 30% of the value we will create. That math works in your favor from month one.
Can we start with just one agent and see how it works?	That is exactly how we recommend starting. Our Starter tier deploys one to two agents. The onboarding investment is $5,000 to $12,000, and the ongoing management is $500 to $1,500 per month. Let us scope that together and identify the single highest-value agent for your business.
We can handle this ourselves.	You can, and you might. The question is whether you want your team spending time on agent configuration, integration maintenance, governance documentation, and performance monitoring -- or whether you want them focused on your clients. We handle the operations. Your team uses the outputs.

Chapter 10 Action Items

1. Practice the business problem opening. In your next three client conversations, open with 'What is the one process that consumes the most time in your operation?' before you mention AI or agents. Track what you learn.

2. Write two client case studies. Document two situations where an agent created measurable business value for a client. Even if these are from early pilots or internal use, quantified results in specific terms are your primary sales evidence.

3. Role-play the four objection responses with your account team until they are natural and confident. A well-handled objection often advances the sale more than a smooth pitch.

CHAPTER ELEVEN

Delivering at Scale Without Burning Out Your Team

The Operations Model That Makes Managed Intelligence Sustainable

Managed Intelligence is a premium service. Clients who pay $4,000 to $8,500 per month for AI agent management have high expectations. They should. Delivering on those expectations consistently, across multiple clients and a growing operations team, requires three things: a standardized delivery model, a scalable operations infrastructure, and a team structured for the service.

The Six Service Commitments

Every Managed Intelligence client receives the same foundational service experience, regardless of engagement size or tenure. Define these commitments in writing, train your team on them, and measure against them every month.

* Monthly AI Performance Report delivered by the 5th business day of the following month, every month.
* 99.5% or better agent uptime across all production agents, measured and reported monthly.
* Incident response within 2 business hours of confirmed detection for any agent impacting client operations.
* Quarterly 90-Day Business Review conducted within the first two weeks of each quarter.
* Governance policy review conducted semi-annually and within 30 days of any material regulatory change.
* Expansion Opportunity List with a minimum of two documented opportunities presented at every 90-Day Review.

The Operations Infrastructure

Three infrastructure investments allow you to run Managed Intelligence at scale without proportional headcount growth.

Centralized Monitoring Platform

A single platform that aggregates performance data from every client's agent environment into a unified dashboard. Your operations team manages the health of your entire Managed Intelligence portfolio from one interface. This platform generates the data that feeds client reports, triggers incident response, and provides the evidence base for optimization recommendations.

Report Automation

Automated report generation that pulls data from your monitoring platform, populates your standard report template, and delivers a draft for human review before client delivery. The goal is to reduce the marginal cost of each monthly report to 30 to 60 minutes of human review time -- not four to six hours of manual data collection and formatting.

Client Portal

A client-facing portal that provides each Managed Intelligence client with continuous access to their agent performance dashboards, historical reports, active governance policies, and expansion opportunity pipeline. Clients who can see their agents performing in real time are more confident in the service and more engaged in expansion conversations. Transparency is your most effective retention tool.

The Four-Role Team

Managed Intelligence Director: Responsible for practice strategy, senior client relationships, and pricing decisions. One director can manage a practice of 15 to 25 active clients.

Agent Operations Manager: Responsible for day-to-day monitoring, incident response, and performance optimization across the client portfolio. One manager can effectively handle 10 to 15 clients at the Professional tier or below.

Agent Engineer: Responsible for agent configuration, integration development, deployment, and technical troubleshooting. One engineer can support four to six active deployments simultaneously.

Client Success Manager: Responsible for client communication, 90-Day Business Review facilitation, reporting delivery, and expansion opportunity identification. One CSM can manage 12 to 18 active Managed Intelligence relationships.

For a practice of 10 clients, four people, one of each role, is financially comfortable given the MRR generated at standard pricing. At 20 clients, add a second Agent Engineer and Client Success Manager. At 30 clients, consider a second Operations Manager.

Chapter 11 Action Items

1. Document your service commitments. Write the six commitments above in a formal service standard document. Share it with every team member and every new client. Measure against it every month.

2. Select your client portal platform. SharePoint, Notion, and dedicated MSP client portal tools are the three most common options. Choose one and build a template client environment before your first production client.

3. Build your Managed Intelligence staffing plan. Based on your 12-month client acquisition target, project when you will need to add each team role. Budget for the investment before you need it.

CHAPTER ELEVEN-B

When Things Go Wrong: The Delivery Recovery Playbook

How to Diagnose and Recover From the Three Failures Every Managed Intelligence Practice Will Face

No delivery model, no matter how well designed, survives contact with real clients without problems. Adoption stalls. An agent accuracy rate comes in below projection. A client's scope expectations diverge from what was scoped. These situations are not signs that Managed Intelligence does not work. They are normal delivery challenges that every practice encounters, and the MSPs who handle them well come out with stronger client relationships than they had before. The ones who are unprepared lose the account.

This chapter gives you the recovery playbook for the three failure scenarios you are most likely to encounter in your first 18 months of Managed Intelligence delivery. Read it before you need it. The decisions you make in the first 48 hours of a delivery problem determine whether it becomes a case study or a cancellation notice.

> **The Most Dangerous Response to a Delivery Problem**
>
> Going quiet. When something is not working, the instinct is often to fix it privately before the client notices. This is the wrong move almost every time. Clients who discover a problem on their own, after your team already knew about it, lose trust in a way that is very difficult to rebuild. Get ahead of problems early. A proactive call to say "here is what we are seeing and here is our plan" almost always lands better than the client raising it first.

Failure Scenario 1: Adoption Failure

Adoption failure is when the agent is technically working, the deployment is complete, the integrations are live, and almost no one on the client's team is using it. The tickets are still being routed manually. The documents are still being processed by hand. The agent sits idle while the team works around it. You are billing for a service that is not creating value because the humans it was built to help are not engaging with it.

Adoption failure is the most common delivery problem in early-stage Managed Intelligence engagements and the one most often misdiagnosed as a technical problem. In almost every case, it is a change management problem. The technology works. The people do not trust it, do not understand it, or were not involved in designing it.

How to Diagnose It

Pull your agent telemetry from Agent 365 for the last 30 days. Look at three numbers: input volume (how many triggers reached the agent), autonomous completion rate (how many it handled without human intervention), and output utilization (how many of its outputs the team acted on rather than overriding). If inputs are low, the workflow was not changed to route work through the agent. If completion rate is low, the instructions need refinement. If output utilization is low, the team is receiving agent outputs but choosing not to use them. That last pattern is the clearest sign of an adoption problem rather than a technical one.

Follow the telemetry with conversations. Talk to two or three team members who were supposed to be using the agent daily. Ask them what their actual workflow looks like. Do not ask leading questions. Listen for phrases like "it is easier to just do it myself," "I do not always trust what it says," or "I did not realize I was supposed to use it for that." Those are your diagnostic signals.

The Recovery Playbook

Adoption Recovery: Steps in Order

→ Stop billing justification, start value demonstration. Schedule a 30-minute session with the team lead and show them the agent processing three real examples live. Seeing it work on their own data is more convincing than any metric you can report.

→ Find and fix the workflow gap. Identify the specific point in the team's daily process where work should enter the agent but is not. In most cases, a simple process change, such as updating a shared inbox rule, a Teams channel routing, or a PSA intake form, is all that is needed. Make the change with them, not for them.

→ Identify and address the trust barrier. Ask the team directly: "What would this agent need to do differently for you to trust its output without checking it?" Their answer is your instruction refinement brief. Make one or two targeted changes to the agent's behavior based on their feedback and show them the result within a week.

→ Assign an internal champion. Ask the client to identify one team member who will own agent adoption on their side. That person becomes your point of contact for adoption issues, attends your monthly check-ins, and advocates internally for the tool. Adoption problems that are owned by someone inside the client organization resolve faster than problems owned entirely by your team.

→ Reset the success criteria conversation. Return to the 30, 60, 90-day benchmarks from the Deployment Blueprint and be honest about where you are. If month two adoption is at 30 percent of target, say so explicitly, explain why, and present a specific plan to reach 70 percent by month three. Clients respond to honesty and a clear plan far better than optimistic status reports.

Pro Tip

The best time to address adoption is before it fails. During the Deploy phase UAT session, ask the process owner: "What would make your team stop using this within 30 days?" Their answer tells you the adoption risk before you go live. Design your go-live plan around mitigating it.

Failure Scenario 2: Accuracy Underperformance

Accuracy underperformance is when the agent is being used but its outputs are consistently wrong, incomplete, or inconsistent in ways that require technician correction. The team is using it, but the correction rate is too high to deliver the efficiency gain you projected. In your ROI model, you projected that the agent would handle 70 percent of inputs autonomously. In practice, it is handling 40 percent and requiring human correction on most of the rest.

This is the scenario most likely to trigger a difficult client conversation about value, because the client is seeing the agent actively in their workflow but not seeing the time savings. They are spending time using the tool and correcting it, which may feel like more work than before, not less.

How to Diagnose It

Pull the agent's error log from Agent 365 for the last 30 days and classify every correction or escalation by type. You are looking for a pattern. If 80 percent of errors involve the same input category or the same decision point, you have a scope gap in the instructions. If errors are distributed randomly across all input types, you likely have a data quality problem in the source systems. If errors cluster around specific times of day or volume spikes, you may have a latency or load issue in the integration layer.

Run 20 of the problematic inputs through the agent in a test environment and observe the reasoning. Copilot Studio's test panel shows you the agent's decision path. Look for where the reasoning diverges from what you intended. In most cases, the agent is not making random errors. It is applying a specific rule incorrectly or encountering an input type that was not covered in the original instruction set.

The Recovery Playbook

Accuracy Recovery: Steps in Order

→ Classify the errors before touching the instructions. Jumping straight to instruction edits without understanding the error pattern typically makes things worse. Spend two hours classifying 30 days of errors before writing a single new line of instruction.

→ Address scope gaps with targeted instruction additions. For each error category you identified, add one specific rule to the instruction set that covers that case explicitly. Test the addition against your error sample before deploying it to production. Measure the impact over the following two weeks.

→ Fix data quality problems at the source. If the issue is in the data feeding the agent, fixing the instructions will not help. Work with the client's team to clean the problematic data source. For recurring data quality issues, add a validation step to the agent's workflow that flags low-quality inputs for human review before processing rather than attempting to process them and producing a bad output.

→ Recalibrate the scope boundary. Some inputs genuinely should not be handled autonomously. If a category of input consistently underperforms despite instruction refinement, add it to the escalation list rather than continuing to attempt automation. A well-scoped agent that handles 80 percent of inputs at 95 percent accuracy is far more valuable than an overscoped agent that attempts everything at 60 percent accuracy.

→ Reset the accuracy benchmark with the client honestly. Show them the error classification you did, explain what you found, and present the specific changes you made. Then set a new 30-day accuracy target together. Clients who understand what caused the problem and see a specific plan to fix it are far more patient than clients who receive vague assurances that things will improve.

Failure Scenario 3: Scope Conflict

Scope conflict is when the client's expectations about what the agent should do differ materially from what was documented in the scope. They believe the agent was supposed to handle a process it was not designed for. Or they believe the agent should be more autonomous than your governance policy permits. Or they are asking for significant changes to the agent's behavior that were not in the original engagement.

Scope conflicts often surface in the 60 to 90 day window, after the client has been living with the agent long enough to form opinions about what they wish it did differently. They are rarely raised as formal complaints. More often, they appear as casual requests in a check-in call: "Can the agent just go ahead and send the invoice without routing it for approval first?" or "We were expecting it to handle the onboarding calls, not just the documents."

The risk with scope conflicts is that the MSP's instinct is to accommodate. Saying yes to scope expansion without pricing it properly, or agreeing to governance changes that compromise the policy framework you built, creates problems that compound over time.

How to Diagnose It

Pull the original scope document and the Governance Policy from the client file. Sit next to the client's request and identify exactly where it diverges from what was scoped. There are three possibilities: the client wants something that was explicitly out of scope (a new capability), the client wants something that was ambiguously scoped (a gray area in the original documentation), or the client believes something was in scope that was never discussed (a misaligned expectation from the sales process). Each of these requires a different response.

The Recovery Playbook

Scope Conflict Recovery: Steps in Order

→ Validate the scope document first, not the client's request. Before responding to any scope request, confirm in writing what the current scope says. Send the client an email that summarizes your understanding of the current scope and asks them to confirm it before the discussion about changes begins. This establishes the baseline without being adversarial.

→ For out-of-scope requests: price it and present it as an expansion. Never accommodate a new capability without a formal change order. Present it as good news: "This is exactly the kind of expansion we built the architecture to support. Here is what adding that capability looks like and what it costs." Use the expansion pricing model from Chapter Nine.

→ For ambiguously scoped items: make a judgment call on the first occurrence, then document it. If something was genuinely ambiguous in the original scope, accommodate the client's interpretation once and then update the scope document to reflect it explicitly. This resolves the ambiguity without creating a precedent for unlimited free changes.

→ For misaligned expectations from the sales process: have the direct conversation. If the client genuinely believed something was in scope that was never offered, acknowledge the misalignment, take ownership of the communication gap, and propose a path forward that is fair to both parties. Do not debate whose fault it was. Resolve it and document the resolution.

→ Never adjust governance policy to resolve a scope conflict. If the client wants the agent to take actions that your governance framework prohibits, the answer is not to change the governance. The answer is to explain why the governance exists and what the client's risk exposure would be without it. If they still want to proceed without the control, put their decision in writing before making any changes.

One pattern runs through all three recovery scenarios: the clients who handle delivery problems best are the ones who were given honest, specific information about what was happening and a clear plan for what comes next. The clients who escalate are almost always the ones who felt they were being managed rather than told the truth. Build your team's communication habits around radical transparency during difficult moments and you will rarely lose an account over a delivery problem.

Chapter 11-B Action Items

1. Before your next client goes live, document what the recovery protocol looks like for each of the three failure scenarios in this chapter. Include who on your team owns the response and what the client communication timeline is. Do this during the Deploy phase, not after a problem occurs.
2. Add an adoption check to your 30-day post-go-live review. Pull the three telemetry metrics: input volume, autonomous completion rate, and output utilization. If any of them is below 60 percent of the projected baseline, initiate the adoption recovery playbook immediately rather than waiting for the monthly report.
3. Review your current scope documents for any active Managed Intelligence clients. Identify any ambiguities, then send a brief email to each client confirming your shared understanding of the current scope. This one proactive step prevents more scope conflicts than any recovery protocol can fix after the fact.

CHAPTER TWELVE

Building the Practice That Defines Your Next Decade

The Roadmap from First Client to Market Leadership

One Managed Intelligence client is a success. Ten is a practice. A practice with documented methodology, a trained team, real case studies, and a growing market reputation is a competitive position that compounds over time. This chapter covers the strategic and organizational work required to build the practice to scale.

Three-Phase Practice Roadmap

Phase 1: Foundation (Months 1 to 6)

Deploy your first two to three Managed Intelligence clients. Focus entirely on delivery excellence. These early engagements test and refine your methodology, build your team's competency, and generate the case studies you need for confident sales conversations. Invest extra internal resource in these clients. The experience and evidence you build here is the foundation of everything that follows.

By month six, you should have two to three live clients, at least one documented case study with quantified business impact, a working monitoring platform and reporting infrastructure, a team confident in five-pillar delivery, and a refined pricing structure based on real-world data.

Phase 2: Scale (Months 7 to 18)

Grow your Managed Intelligence client base using the case studies and referrals from Phase 1. Launch a systematic outreach program targeting your highest-potential existing clients. Develop vertical-specific agent use case libraries for the two industries you serve most deeply. Invest in practice infrastructure -- monitoring platform maturation, report automation, client portal launch, and team expansion.

By month 18, target 10 to 20 active Managed Intelligence clients, $50,000 to $150,000 in monthly Managed Intelligence MRR, and a recognized practice identity in your market.

Phase 3: Leadership (Months 19 to 36)

Establish category leadership. Publish thought leadership content -- case studies, conference presentations, white papers, and potentially an annual State of MSP AI Intelligence report. Explore strategic acquisitions of small, technically specialized AI boutiques that accelerate your capability. Consider a white-label or channel partner model that allows other MSPs to deliver Managed Intelligence under your methodology.

By month 36, your Managed Intelligence practice should be generating $2 million to $5 million in annual recurring revenue and your position in the market should be one that no late-entering competitor can displace quickly.

Vertical Specialization

The MSPs who build the largest and most defensible Managed Intelligence practices choose a vertical and go deep. Vertical specialization lets you develop reusable use case libraries, industry-specific governance frameworks, and ROI benchmarks that make every subsequent engagement faster, cheaper, and more credible.

Vertical	Opportunity Landscape and Key Differentiator
Healthcare	Patient communication, clinical documentation, prior authorization, appointment optimization. HIPAA governance expertise is your primary differentiator.
Professional Services (Legal, Accounting, Consulting)	Document review, contract abstraction, client onboarding, billing capture. Confidentiality governance and privilege protection are the key differentiators.
Financial Services	Client onboarding, KYC processing, compliance reporting, transaction monitoring. SOC 2 and FINRA compliance expertise is what wins deals.
Manufacturing and Distribution	Purchase order processing, supplier communication, inventory forecasting, quality control documentation. ERP integration expertise is your edge.
Real Estate	Listing generation, client communication, lease abstraction, market analysis. Data accuracy and fair housing compliance are the governance considerations.

Three Essential Partnership Categories

Platform Partnerships: Deep relationships with Microsoft for Copilot Studio, OpenAI for the Agents SDK, and your workflow automation platform of choice. These partnerships provide technical enablement, co-marketing, and early access to new capabilities.

Delivery Partnerships: Relationships with specialized AI engineering firms who can execute the technically complex deployments that exceed your current internal capability. These partners let you scope and sell engagements you could not deliver alone while building the internal capability over time.

Compliance Partnerships: Relationships with regulatory compliance experts and legal advisors who support your governance practice in specific regulatory domains. These are particularly valuable for healthcare and financial services clients.

Chapter 12 Action Items

1. Write your Managed Intelligence practice charter. Define your target market, core value proposition, service scope, pricing strategy, and 18-month revenue target. This document is your practice-building north star.
2. Choose your vertical specialization. Based on your existing client portfolio, select the one vertical where you will build the deepest expertise in the next 12 months. Document the top five agent use cases and primary governance requirements for that vertical.
3. Start your platform partnership conversation. Contact your Microsoft Partner team about Copilot Studio enablement resources. This week. Not next quarter.

Conclusion: The MSPs Who Act Now Will Own the Market

Thirty-five years in managed services teaches you what distinguishes the technology shifts that reshape the industry from the ones that generate noise for 18 months and then fade. Cloud reshaped the industry. Cybersecurity as a managed service reshaped the industry. Both created multi-year windows during which the early movers established market positions their later-entering competitors never dislodged.

AI agent management will reshape the industry. The window is open right now, in 2026. The gap between what your clients need and what the market currently provides is wide enough to build a substantial practice before competition catches up.

This book gives you the framework. The pricing model gives you the commercial structure. The delivery methodology gives you a repeatable service. The governance framework gives you the differentiator. What remains is your decision to act.

Start with one client. Complete the Discover engagement. Build one agent. Monitor it professionally. Report the results. Present the first expansion opportunity. Let the compounding revenue begin. The first client is the hardest. By the tenth, you have a practice generating extraordinary recurring revenue with client relationships that are deeper and more defensible than anything traditional managed services has ever produced.

> *"The clients who will matter most to your business in 2030 are the ones you start operating AI for in 2026. Build the practice. Deliver with excellence. Own the category."*
> **-- Tony Ferrigno**

Tony Ferrigno
Founder and Managing Partner, AiT Advisory Group
We Put AI in IT

Appendix A: Service Tiers and Pricing Reference

Use these ranges as your starting point. Adjust based on your market, your cost structure, and the specific complexity of each engagement. All recurring fees assume a 24-month minimum commitment. Month-to-month agreements carry a 25% premium.

Onboarding Investment

Starter	Standard	Advanced	Enterprise
1 to 2 agents, 2 to 3 system integrations	3 to 5 agents, 4 to 6 system integrations	6 to 10 agents, complex integrations	10 or more agents, orchestration tier
$5,000 to $12,000	$12,000 to $28,000	$28,000 to $60,000	$60,000 to $150,000 or more
3 to 4 weeks	4 to 6 weeks	6 to 10 weeks	10 to 20 weeks

Monthly Management Fee

Foundation	Essentials	Professional	Enterprise
1 to 3 agents	4 to 8 agents	9 to 15 agents	16 or more agents
$500 to $1,500/month	$1,500 to $4,000/month	$4,000 to $8,500/month	$8,500 to $20,000 or more/month
$6,000 to $18,000/year	$18,000 to $48,000/year	$48,000 to $102,000/year	$102,000 to $240,000 or more/year

Expansion Agent Pricing

Expansion Type	Pricing
Additional agent using existing integration infrastructure	60 to 70% of standard onboarding rate for equivalent complexity
Additional agent requiring new system integration	80 to 90% of standard onboarding rate for equivalent complexity
Tier upgrade with no new agents	Difference in monthly rate only; no setup fee
Standalone Governance Retainer	$800 to $2,500 per month depending on regulatory complexity and audit requirements

Appendix B: Agent Use Case Library by Vertical

Use this library to populate your Expansion Backlog and to prepare for vertical-specific client conversations. Each use case includes an estimated business impact, the primary platform, and a monthly management range. Impact estimates are based on documented benchmarks from production deployments.

Cross-Vertical Use Cases

Use Case	Description and Impact	Platform	Monthly Management
Helpdesk Ticket Triage and Routing	Classify, prioritize, and route incoming support tickets. 60 to 70% autonomous resolution for Tier 1 issues in production deployments.	Copilot Studio, ServiceNow AI	$800 to $2,000/month
Meeting Transcription and Action Items	Transcribe meetings, extract action items, and distribute summaries to relevant stakeholders automatically.	Microsoft Teams Copilot	$300 to $600/month
Document Summarization and Review	Summarize long documents, extract key information, and flag items requiring human attention.	OpenAI API, Copilot Studio	$400 to $900/month
Email Triage and Response Drafting	Classify incoming email, draft responses to routine inquiries, escalate complex items to human review.	Copilot for Outlook, Zapier AI	$500 to $1,200/month
Compliance Report Generation	Compile data from multiple systems and generate standard compliance reports on a defined schedule.	Power Automate with AI Builder	$600 to $1,500/month
Vendor Invoice Processing	Extract data from vendor invoices, match to purchase orders, flag discrepancies, and route for approval.	Document AI, Power Automate	$700 to $1,800/month

Healthcare

Use Case	Description and Impact	Platform	Monthly Management
Patient Communication and Scheduling	Handle routine patient inquiries, appointment scheduling, and reminder communications. Reduces front desk call volume by 40 to 60%.	Healthcare CRM with Copilot Studio	$1,200 to $3,000/month
Prior Authorization Processing	Extract required information from clinical documentation and prepare prior authorization submissions. Cuts processing time by 65% on average.	Document AI with EHR integration	$1,500 to $4,000/month
Clinical Note Transcription	Transcribe and structure physician notes from voice recordings into EHR-formatted documentation.	Azure Speech with Document AI	$800 to $2,000/month
Patient Intake Automation	Collect and validate patient and insurance information, identify coverage gaps, and prepare intake packets.	Copilot Studio with EHR integration	$1,000 to $2,500/month

Professional Services

Use Case	Description and Impact	Platform	Monthly Management
Contract Review and Abstraction	Review contracts for key terms, flag unusual clauses, and extract structured data for matter management. Reduces attorney review time by 50 to 70% on standard contracts.	OpenAI API with DMS integration	$1,200 to $3,500/month
Client Onboarding Automation	Collect engagement information, run conflict checks, generate engagement letters, create client records.	Copilot Studio with practice management	$800 to $2,000/month
Time Entry and Billing Capture	Review work descriptions and communications, suggest time entries, and flag billing opportunities missed in daily capture.	OpenAI API with billing system	$600 to $1,500/month
Research Summary Generation	Compile and summarize case law, regulatory guidance, or	OpenAI API with legal research tools	$900 to $2,500/month

Use Case	Description and Impact	Platform	Monthly Management
	market intelligence from defined sources.		

Financial Services

Use Case	Description and Impact	Platform	Monthly Management
Client Onboarding and KYC Processing	Collect and validate client identity information, run KYC checks, and prepare compliance documentation. Reduces onboarding cycle from 5 days to under 24 hours in documented deployments.	Document AI with CRM	$1,500 to $4,000/month
Financial Plan Generation Support	Compile client data, populate plan sections from templates, and prepare advisor review packages.	OpenAI API with planning software	$1,000 to $2,500/month
Transaction Anomaly Detection	Monitor patterns, flag anomalies for compliance review, and generate initial SAR documentation drafts.	AI Builder with core banking	$2,000 to $5,000/month
Regulatory Filing Preparation	Compile required data from multiple systems and prepare structured regulatory filing packages on schedule.	Power Automate with compliance platform	$1,200 to $3,000/month

Appendix C: Governance Policy Framework Template

Complete one Governance Policy Document for each deployed agent. This is a living document. Review it semi-annually and update it whenever the agent's scope, connected systems, or applicable regulatory environment changes materially.

SECTION 1: AGENT IDENTIFICATION

Field	Value
Agent Name	
Agent ID	Convention: [CLIENT]-[FUNCTION]-[NUMBER]. Example: ACME-HELPDESK-001
Client Organization	
Primary Function	One sentence: what does this agent do?
Platform	e.g., Microsoft Copilot Studio, OpenAI Agents SDK, Make
Deployment Date	
Policy Version	
MSP Policy Owner	Name and role of the MSP team member responsible for this policy
Client Approver	Name and role of the client executive who approved this policy
Next Review Date	

SECTION 2: DATA GOVERNANCE

Element	Policy
Data Sources Accessed	All systems and data sources this agent reads from
Data Types Processed	All categories of data the agent reads, processes, or generates
Sensitive Data Classification	Any PII, PHI, financial, or confidential data within scope
Data Retention	How long are agent inputs, outputs, and logs retained?
Data Residency	Where is data processed and stored? Geographic restrictions?
Third-Party Data Sharing	Does the agent send data to any external system?
Access Controls	Who can access agent data, logs, and configuration?

SECTION 3: OPERATIONAL GUARDRAILS

Guardrail Category	Policy
Permitted Autonomous Actions	Explicit list of actions this agent can take without human approval
Prohibited Actions	Explicit list of actions this agent cannot take under any circumstances
Human Approval Required For	Action types that require human review before execution
Escalation Triggers	Conditions that cause the agent to pause and route to a human reviewer
Communication Authorization	What communications can this agent send autonomously? To whom?
Scope Boundaries	The business processes within this agent's defined operational scope
Maximum Transaction Value	For agents initiating financial transactions, the ceiling without human approval

SECTION 4: COMPLIANCE AND REGULATORY

Element	Policy
Applicable Regulations	All regulatory frameworks that apply to this agent's operation
Compliance Controls	Specific technical and operational controls for each applicable regulation
Audit Log Requirements	What must be logged, at what granularity, and for how long?
Regulatory Change Protocol	How is this policy updated when applicable regulations change?
External Audit Support	What documentation and access can be provided for regulatory examination?

SECTION 5: INCIDENT RESPONSE

Element	Protocol
Incident Definition	What constitutes a reportable incident for this specific agent?
Detection Method	How will incidents be detected? (Alert, client report, routine audit)
Initial Response SLA	Maximum time from detection to initial response: ___ hours
Escalation Path	Who is notified at each severity level?
Client Notification SLA	Maximum time from confirmed incident to client notification: ___ hours
Containment Protocol	What action is taken immediately to limit scope?
Post-Incident Review	What governance policy updates result from a confirmed incident?

Appendix D: Agent Operations Runbook Template

The Agent Operations Runbook is your team's field guide for managing a specific deployed agent. Every production agent in your Managed Intelligence portfolio should have a current Runbook maintained in your operations management system and reviewed quarterly.

RUNBOOK HEADER

Field	Value
Agent Name and ID	
Client	
Runbook Version	
Primary Operations Owner	
Backup Operations Owner	
Client Primary Contact	
Client Escalation Contact	
Last Updated	
Platform	
Monitoring Dashboard URL	

SECTION 1: AGENT OVERVIEW

Element	Value
Primary Function	What does this agent do?
Business Process Owner	The client-side person responsible for this process
Connected Systems	All integrated platforms and their connection type
Input Sources	What triggers agent action?
Output Destinations	Where does the agent send its outputs?
Expected Daily Processing Volume	Inputs per day -- used as baseline for volume anomaly alerting
SLA Commitments	Response time, completion time, and uptime requirements
Operating Hours	24/7 or business hours only?

SECTION 2: NORMAL OPERATIONS PROCEDURES

Procedure	Detail
Daily Health Check	Log into monitoring dashboard. Verify: (1) Agent status is Active, (2) No open alerts, (3) Processing volume within 20% of daily baseline, (4) All integration connections show Green. If all clear, log the check as complete. If any item fails, proceed to Alert Response in Section 3.
Weekly Performance Review	Review 7-day performance metrics: task completion rate, accuracy rate, throughput, and SLA compliance. Flag any metric deviating more than 10% from the prior 4-week average. Document findings in the operations log. Escalate to optimization review if the deviation persists for more than two weeks.
Monthly Report Preparation	Pull reporting period data from the monitoring platform. Populate the Monthly AI Performance Report template. Calculate business impact metrics: hours saved, errors eliminated, and cost savings. Write the optimization recommendations section based on the month's data. Submit draft to the Managed Intelligence Director for review by the 3rd business day of the following month.

SECTION 3: ALERT RESPONSE PROCEDURES

Alert Level	Response Procedure
P1 -- Agent Offline (uptime = 0%)	IMMEDIATE. Contact client within 30 minutes. Investigate root cause: integration failure, platform outage, or authentication expiry. Initiate platform vendor escalation if required. Document in incident log. Target restoration: 2 hours.
P2 -- Integration Failure (one or more connections down)	High. Investigate within 1 hour. Attempt reconnection. If unresolved in 2 hours, escalate to Agent Engineer. Notify client if business operations are affected. Document in incident log.
P3 -- Performance Degradation (completion rate below 80% of baseline)	Medium. Investigate error logs within 4 hours. Identify root cause: model behavior change, data quality issue, or scope problem. Implement fix. Monitor for 24 hours after resolution. Include in monthly report.
P4 -- Anomalous Output Detected	Medium. Pause agent processing if outputs are being sent to external parties. Review anomalous outputs. Determine whether a governance policy violation occurred. If yes, initiate Incident Response per Section 5 of the Governance Policy. If no, adjust agent instructions and resume.
P5 -- Volume Anomaly (more than 50% deviation from baseline)	Low. Investigate cause: seasonal pattern, process change, or data source issue. Confirm with client whether a business change occurred. Document and monitor. No immediate action unless output quality is also affected.

SECTION 4: MAINTENANCE SCHEDULE

Cadence	Activities
Weekly	Integration health check, performance metric review, error log analysis
Monthly	Report generation, optimization review, SLA compliance confirmation
Quarterly	Governance policy review, model performance audit, expansion opportunity assessment, 90-Day Business Review preparation
Semi-Annual	Governance policy update for regulatory changes, full integration compatibility review, disaster recovery test
Annual	Full agent architecture review, technology refresh assessment, SLA review

SECTION 5: ESCALATION CONTACTS

Contact Role	Information
MSP Operations Team	Phone: ___ Email: ___ Hours: ___
MSP Agent Engineer	Phone: ___ Email: ___
MSP Managed Intelligence Director	Phone: ___ Email: ___
Client Primary Contact	Phone: ___ Email: ___
Client Executive Escalation	Phone: ___ Email: ___
Platform Vendor Support	Support URL: ___ Priority support line: ___

Appendix E: Sample Agent Opportunity Report

The Agent Opportunity Report is the primary deliverable of the Discover phase. The sample below is completed for a fictional client, Meridian Accounting Group, a 38-person CPA firm with three office locations. Use it as your formatting template and as a calibration reference for scoring, ROI modeling, and recommendation structure. Every field shown here should appear in your actual client reports. The numbers and findings will differ. The structure should not.

Client Snapshot

Client	Meridian Accounting Group
Industry	Public Accounting / CPA Firm
Staff Count	38 employees across 3 locations
Annual Revenue	Approximately $4.2 million
Primary Systems	Microsoft 365 E3, QuickBooks Online, Thomson Reuters UltraTax, Karbon (practice management)
Compliance Context	AICPA standards, IRS e-filing requirements, state CPA board regulations
Assessment Date	[Date]
Prepared By	[MSP Name] Managed Intelligence Practice

Executive Summary

Meridian Accounting Group operates a high-volume, knowledge-work-intensive practice with significant manual process burden concentrated in three operational areas: client onboarding and engagement setup, tax document collection and processing, and client communication management. The assessment identified four Tier 1 agent opportunities with a combined estimated annual value of $214,000 in recovered staff capacity. Three of the four opportunities are deployable within the current Microsoft 365 E3 licensing without additional platform investment.

The firm's data environment is well-structured and Microsoft-aligned. Primary integration readiness gaps exist in the QuickBooks Online and Thomson Reuters UltraTax connections, both of which are addressable through existing API capabilities with an estimated 8 to 12 hours of integration configuration work per platform. Workforce readiness is strong among administrative staff and moderate among professional staff, with targeted change management recommended for the senior associate group.

Process Inventory and Priority Scoring

Each candidate process was scored on four dimensions using a 1 to 5 scale. Tier 1 (total score 16 to 20) indicates immediate deployment candidates. Tier 2 (10 to 15) indicates second-phase candidates. Below 10 indicates deprioritized.

Process	Frequency	Manual Effort	Error Rate	Agent Suitability	Score
Client onboarding and engagement setup	5	5	4	5	**19 / Tier 1**
Tax document collection and tracking	5	5	3	5	**18 / Tier 1**
Client email triage and response drafting	5	4	3	4	**16 / Tier 1**
Time entry capture and billing review	4	4	4	4	**16 / Tier 1**
Research summary generation	3	4	2	4	**13 / Tier 2**
Staff scheduling and deadline tracking	3	3	2	3	**11 / Tier 2**
Strategic client planning and advisory	2	2	1	2	**7 / Tier 3**

ROI Model: Tier 1 Opportunities

The model below uses a fully loaded staff cost of $52 per hour (blended rate across administrative and professional staff based on client-provided payroll data) and a conservative 60 percent reduction in manual time for each automated process. Actual results typically range from 55 to 75 percent depending on data quality and process complexity.

Process / Agent	Hrs/Week (Current)	Monthly Cost (Current)	Monthly Savings (60%)	Annual Value
Client onboarding agent	14 hrs	$2,912	$1,747	$20,966
Document collection agent	18 hrs	$3,744	$2,246	$26,957
Email triage agent	12 hrs	$2,496	$1,498	$17,971
Time entry and billing agent	11 hrs	$2,288	$1,373	$16,474
Total (4 agents)	**55 hrs/week**	**$11,440/month**	**$6,864/month**	**$82,368 in recovered annual value**

Deployment Blueprint

The recommended deployment sequence prioritizes the client onboarding agent first because it produces visible results quickly, affects the entire staff, and sets the adoption tone for subsequent deployments. The document collection agent deploys second because it builds on the same Microsoft 365 integration foundation. The email triage and time entry agents deploy in months three and four respectively.

Phase 1 (Weeks 1-6): Client onboarding agent. Platform: Microsoft Copilot Studio. Integrations: Microsoft 365, Karbon practice management API. Governance tier: Standard. Estimated management fee: $1,200 per month.

Phase 2 (Weeks 5-10): Tax document collection agent. Platform: Microsoft Copilot Studio with SharePoint integration. Integrations: Microsoft 365, SharePoint, client email domain. Governance tier: Standard with AICPA compliance documentation. Estimated management fee: $1,400 per month.

Phase 3 (Weeks 9-14): Email triage agent. Platform: Microsoft Copilot for Outlook with Copilot Studio orchestration. Integrations: Microsoft 365, Karbon. Governance tier: Standard. Estimated management fee: $900 per month.

Phase 4 (Weeks 13-18): Time entry and billing agent. Platform: Microsoft Copilot Studio. Integrations: Microsoft 365, QuickBooks Online, Karbon. Governance tier: Standard with billing audit trail. Estimated management fee: $1,100 per month.

Engagement Investment Summary

Onboarding investment (Discover + 4-agent Deploy): $22,500 fixed fee
Monthly management fee at full deployment (Standard tier, 4 agents): $4,600 per month
Year 1 total investment: $22,500 + ($4,600 x 8 months active) = $59,300
Year 1 value created (conservative 60% savings): $82,368
Net Year 1 benefit: $23,068 after full investment recovery
Year 2 ongoing investment: $55,200 | Year 2 value: $82,368+ (expanding with optimization)
Breakeven: Month 4 of active management

Appendix F: Managed Intelligence Sales Conversation Scripts

Chapter Ten gives you the framework for the Managed Intelligence sales conversation. These scripts give you the actual language. Every script here has been structured around the principle that the best sales conversation is a diagnostic conversation: you ask, the client talks, you listen, and the solution you propose is specific to what you heard.

Use these scripts as starting points, not as recitations. The language should sound like you. Read each script out loud before you use it, adapt the phrasing to your natural speaking style, and customize the industry references and examples to match your prospect. A script that sounds scripted loses the trust you are trying to build.

Script 1: Opening an AI Conversation With an Existing Client

Use this when you are transitioning an established managed services client into a Managed Intelligence conversation for the first time. The goal is to open the door without creating pressure. You are planting a seed, not closing a deal.

Conversation Opener

[Client name], I want to share something we've been seeing across our client base that I think is directly relevant to you.

Over the last six months, we've been deploying AI agents for a handful of our clients -- software that takes over specific, repetitive processes their teams were doing manually. The results have been significant. One client recovered 14 hours of staff time per week in the first 60 days.

What I want to ask you is this: if you could take one process that consumes a lot of your team's time and have it handled automatically and accurately, with no manual involvement, what would that process be?

[Let the client answer. Do not interrupt. Whatever they name is your entry point.]

Tell me more about that process. How often does it happen? Who's involved? What does it look like when it goes wrong?

[Listen fully. Take notes. This is your Discovery conversation beginning.]

Here's what I'd like to propose. We do a structured assessment -- a paid engagement, typically four to six weeks -- where we map out exactly which of your processes are the best candidates for AI automation, model the value each one creates, and give you a detailed deployment plan. It's the foundation of everything we'd build together. Would you like to hear what that looks like?

Script 2: The Business Pain Discovery Sequence

Use this sequence in the first substantive meeting after the client has expressed interest. Your goal is to quantify the pain, not describe the technology. Every question moves toward a specific number.

Discovery Sequence (in order)

QUESTION 1 -- The volume question: How many times per week does [the process they named] happen? Roughly.

[After their answer]: And how many people are involved each time it runs?

QUESTION 2 -- The time question: If you had to estimate, how many minutes does it take one person to complete each instance of that process from start to finish? Including all the back and forth?

[Calculate mentally: instances per week x people x minutes. You will need this number.]

QUESTION 3 -- The error question: What happens when it goes wrong? How often does that happen? What does it cost you when it does?

[This surfaces the error rate dimension and often reveals a secondary cost the initial calculation missed.]

QUESTION 4 -- The frustration question: On a scale of 1 to 10, how much does this process frustrate your team? What do they say about it?

[This is a change management question dressed as a diagnostic question. A 7 or higher signals strong adoption readiness when you solve it.]

QUESTION 5 -- The priority question: If this process were handled automatically, accurately, every time, without anyone on your team touching it -- what would they do with that time?

[The answer to this question is your value story. Use it in your proposal.]

Script 3: Proposing the Assessment Engagement

Use this immediately after a productive Discovery conversation while the pain is vivid. Proposing the assessment in the same meeting as the discovery conversation increases conversion by 40 to 60 percent versus following up later.

Assessment Proposal Close

Based on what you've just described, I can see at least two, possibly three processes where AI agents would create significant impact for your team. The onboarding process and the [second process they mentioned] are the ones I'd want to look at most closely.

Here's what I'd recommend as a next step. We do what we call a Discover engagement. It's a structured, paid assessment where we go through your operations systematically, score every candidate process against four objective criteria, model the ROI for the top ones, and build you a deployment blueprint. You walk away with a written plan for exactly which agents we'd build, in what sequence, and what each one is worth.

For an organization your size, that engagement is [price]. It takes four to six weeks. And it's the foundation of everything we'd build together after that.

The reason I'm recommending you start here rather than just jumping into building something is that the assessment consistently surfaces opportunities that aren't obvious at the start. For one client, we found a billing process that was consuming 22 hours per week that nobody had flagged as a problem -- they just thought it was how billing worked.

Does it make sense to set up a scope call this week to outline exactly what the assessment would cover for your business?

Script 4: Handling "We Can Handle This Internally"

This is the most common serious objection to Managed Intelligence and the one most likely to be sincere rather than a negotiating position. Handle it with respect, not dismissal. The goal is not to argue that they cannot handle it -- it is to surface what handling it actually requires.

Full Objection Response

That's a legitimate path, and some organizations do manage their AI agents internally. I want to give you an honest picture of what that involves so you can make the right call for your business.

Managing AI agents well is not the same as deploying them. Deployment is a project -- it has a start and an end. Management is ongoing. It includes daily health monitoring, performance analytics, integration maintenance as your connected systems update, governance documentation, audit trails, incident response when an agent behaves unexpectedly, and quarterly reviews to identify what to build next.

The question isn't whether your team is capable. I'm sure they are. The question is whether you want them spending time on agent operations -- or whether you want them focused on your clients and your core business.

Let me ask you something directly: who on your team would own the agent operations function? What does their current workload look like? And do they have experience with the specific monitoring and governance work that makes agent management reliable?

[Let them answer. The answer usually reveals the gap.]

What we provide is a team that does only this. We have documented methodologies, monitoring infrastructure, incident response protocols, and reporting systems that took us 18 months to build. You'd be building that from scratch while also running your business. We can talk about what that looks like -- or we can talk about what it would look like to have that capability available to you from day one.

Which conversation would be more useful right now?

Script 5: The 90-Day Review Expansion Opener

Use this at the opening of every 90-Day Business Review to shift the conversation from status update to strategic growth. The expansion conversation should feel like an advisory moment, not a sales pitch. The difference is that you are presenting solutions to problems the client already knows they have.

90-Day Review Expansion Opener

Before we look at the performance numbers, I want to share something we've been tracking over the last 90 days that I think will be relevant to the second part of our conversation today.

While we've been running your [Agent Name] and [Agent Name], our operations team has been paying attention to the broader operational context -- the processes your team is still handling manually, the exceptions that come up repeatedly in the agent logs, and the workflow patterns that suggest where the next opportunity is.

I've documented three specific observations that I want your input on before we talk about next steps. None of this is a sales pitch -- it's just what we've been seeing from our position inside your operation.

[Present Observation 1 -- a specific process you noticed consuming manual time, with an estimate of the volume.]

[Present Observation 2 -- an exception pattern from the agent logs that suggests a scope expansion opportunity.]

[Present Observation 3 -- a workflow you've identified through your broader IT visibility that aligns with the client's stated growth priorities.]

I have initial thinking on what an agent solution for each of these would look like, including rough value estimates. But before I go there, I want to know -- do any of these resonate with what you've been experiencing?

[Let them confirm or redirect. Their reaction tells you which opportunity to prioritize.]

Script 6: Closing the Pilot Proposal

Use this when a prospect wants to start but is not ready to commit to a full Standard or Advanced engagement. A well-scoped pilot converts to a full engagement in over 80 percent of cases when the agent delivers on its defined success criteria. Frame the pilot as a structured commitment, not a discounted trial.

Pilot Proposal Close

I understand you want to see it work before you commit to the full program. That's a reasonable position, and it's actually exactly how we recommend starting.

Here's what a pilot looks like with us. We don't do open-ended trials. We structure the pilot as a defined project: one specific agent, one specific process, with agreed-upon success criteria that you and I define together before we start. At 90 days, we measure the actual results against those criteria and decide together what comes next.

For your situation, the process I'd recommend for the pilot is [specific process]. It's high-frequency, it has measurable output, and it's representative of the broader opportunity -- so the results we get from it will tell you a lot about what the full program would deliver.

The pilot investment is [price]. That covers the Discover work specific to this process, the full deployment, and 90 days of active management and reporting. At the end of 90 days, you'll have a live agent producing measurable results, a documented performance report, and a clear picture of what the next phase looks like.

You're not committing to anything beyond the pilot. But in my experience, when the agent does what we've projected -- and they do, when the process is right and the deployment is done correctly -- the decision about what to do next is easy.

Can we define the success criteria together right now? That's the starting point.

A note on using these scripts in practice: the most effective sales conversations are the ones where the client talks more than you do. Every script above is structured to generate client responses, not to fill silence with your own voice. Ask the question. Stop talking. The discomfort of silence is yours, not the client's. Resist the urge to fill it.

Appendix G: The Managed Intelligence 90-Day Launch Checklist

This checklist consolidates the action items from every chapter into a single week-by-week sequence for your first 90 days. Use it as your launch plan. Each item references the chapter where the full context lives. Print it, pin it, and work through it with your team. The MSPs who move from reading to executing in under two weeks capture the early-mover advantage. The ones who wait until everything feels perfect typically wait too long.

This checklist assumes you have read the book. It is not a summary of the concepts. It is a sequenced list of concrete actions with deadlines and chapter references so you can find the detail when you need it.

PHASE 1: INTERNAL READINESS | Weeks 1 and 2

Everything in Phase 1 happens inside your MSP before you have a single client conversation. The goal is to be ready to demo, scope, and propose before you talk to anyone. MSPs who approach their first prospect without this foundation in place lose credibility in the first meeting.

Ch. 2-B	☐ Confirm Microsoft 365 licensing tier for your MSP tenant supports Copilot Studio and Agent 365
Ch. 2-B	☐ Enable Agent 365 in your own MSP tenant and run the discovery scan
Ch. 2-C	☐ Confirm Power Platform environment is configured with correct admin roles
Ch. 2-B	☐ Set up delegated admin access to the first three target client tenants

Ch. 2-C	☐ Build your first internal agent: helpdesk ticket triage for your own support inbox
Ch. 6	☐ Select your centralized monitoring platform (Azure Monitor, Datadog, or equivalent)
Ch. 2-B	☐ Connect your MSP tenant to Agent 365 telemetry and confirm data flow
Ch. 6	☐ Build your Monthly AI Performance Report template in its final format
Ch. 11	☐ Select your client portal platform (SharePoint, Notion, or dedicated tool)

Ch. 11	☐ Assign the four Managed Intelligence roles: Director, Ops Manager, Agent Engineer, CSM
Ch. 2	☐ Complete Microsoft Copilot Studio learning path for your Agent Engineer
Ch. 2-B	☐ Complete Agent 365 admin training for your Ops Manager
Ch. 2-C	☐ Run your team through the internal agent you built: explain what it does and how to manage it

Ch. 11	☐ Write and distribute your six service commitment standards to every team member

Ch. 9	☐ Build your Managed Intelligence pricing calculator with all four tiers and commitment discounts
Ch. 9	☐ Set your price floors in writing and share with anyone authorized to quote
Ch. 4	☐ Write your Discover engagement proposal template (scope, deliverables, price)
Ch. 4	☐ Build your ROI model template with the four standard input fields
Ch. 10	☐ Practice the business problem opening question out loud five times before your first prospect call

PHASE 2: FIRST PROSPECT | Weeks 3 through 6

Phase 2 is your first real client engagement. Everything in this phase should happen with one specific prospect: the client you identified in Chapter One as the most operationally mature, highest-relationship, and highest-process-volume account in your portfolio. Do not attempt to run three discovery conversations simultaneously in your first 30 days. Go deep on one.

Ch. 1	☐ Identify your first Managed Intelligence prospect by applying the three criteria: process volume, relationship strength, operational maturity
Ch. 2-B	☐ Enable Agent 365 in their tenant and run the discovery scan before the first conversation
Ch. 2	☐ Review their current Microsoft 365 usage and identify agent capabilities already available to them
Ch. 2-B	☐ Prepare two or three specific process observations from the Agent 365 scan to open the conversation

Ch. 10 / App. F Script 1	☐ Open with the business problem question, not an AI pitch
Ch. 10 / App. F Script 2	☐ Run the five-question discovery sequence and record the answers with specific numbers
Ch. 4	☐ Calculate the manual cost of the process they described before leaving the meeting
Ch. 10	☐ Send the discovery meeting follow-up email within 24 hours with the outline you discussed
Ch. 10 / App. F Script 4	☐ Handle any objections using the four response frameworks before the meeting ends

Ch. 10 / App. F Script 3	☐ Propose the Discover engagement at the close of the discovery conversation, not in a follow-up email

Ch. 4	☐ Deliver the signed Discover engagement agreement within 48 hours of verbal approval
Ch. 4	☐ Complete the five-dimension assessment: process inventory, data readiness, compliance, workforce readiness, ROI model
Ch. 4 / App. E	☐ Deliver the Agent Opportunity Report and Deployment Blueprint on the committed date
Ch. 4	☐ Present the findings with the ROI model prominently featured in the first five minutes
Ch. 10	☐ Close the Deploy engagement at the Discover presentation meeting

PHASE 3: FIRST DEPLOYMENT AND GO-LIVE | Weeks 7 through 12

Phase 3 is execution. By Week 7 you have a signed engagement, a completed assessment, and a client ready to go. This phase follows the Deploy pillar methodology exactly. The timeline below assumes a Starter or Standard engagement (one to three agents). Larger engagements run longer but follow the same sequence.

Ch. 2-C	☐ Complete the pre-build checklist: service accounts, permissions, scope document, test data
Ch. 5 / Ch. 2-C Step 3	☐ Build all integration connections and test each one independently before agent configuration
Ch. 2-C Step 2	☐ Configure the agent instructions covering all five sections: role, classification, rules, routing, escalations
Ch. 2-C Step 4	☐ Configure guardrails, content moderation, authentication, and fallback topic
Ch. 2-B	☐ Confirm Agent 365 is receiving telemetry for the new agent before proceeding to testing

Ch. 2-C Step 5	☐ Run full test library of 20 or more sample inputs and document results in a test log
Ch. 2-C Step 5	☐ Calculate autonomous accuracy rate: must be above 80 percent before UAT
Ch. 5 / Ch. 2-C Step 5	☐ Complete UAT session with client process owner and incorporate every correction
Ch. 7 / App. C	☐ Add the Governance Policy document to the client file before go-live
App. D	☐ Complete the Operations Runbook for this agent

Ch. 2-C Step 6	☐ Publish agent and restrict to pilot group for first 30 days
Ch. 2-C Step 6	☐ Run live end-to-end test: confirm PSA ticket, acknowledgment email, and Agent 365 log

Ch. 2-C Step 6	☐ Notify client team of go-live with expected response times and escalation contact
Ch. 6	☐ Review agent telemetry daily for the first two weeks: flag any deviation from baseline immediately
Ch. 11-B	☐ Conduct 30-day adoption check: input volume, autonomous completion rate, output utilization
Ch. 6	☐ Deliver Month 1 AI Performance Report by the 5th business day of the following month
Ch. 8	☐ Schedule the 90-Day Business Review before leaving the go-live meeting
Ch. 8	☐ Begin populating the Expansion Backlog based on operational observations in the first 30 days

Pro Tip

The single most common reason MSPs stall between reading a book like this and building the practice is the feeling that they need to complete every item on a list like this before talking to their first prospect. You do not. Complete Phase 1 fully. Then start the prospect conversation in parallel with Phase 2. The discovery conversation will sharpen everything you are still building. Done is better than perfect when the window is open.

Track your progress against this checklist weekly. Share it with your Managed Intelligence Director and review it in your Monday operations meeting. At the 90-day mark, you should have a live client, a delivered Month 1 report, and a scheduled 90-Day Business Review. Everything in this book was written to get you to that point. The rest builds from there.

About AiT Advisory Group™

AiT Advisory Group™ is a specialized AI and IT consulting firm focused exclusively on helping Managed Service Providers monetize, deploy, and scale artificial intelligence and other managed services. Founded by Tony Ferrigno, a 35-year MSP industry veteran, AiT Advisory Group brings the most relevant and practical AI guidance available to MSPs across national, regional, and global markets.

Our signature tagline, 'We Put AI in IT™,' captures our mission precisely. We work at the intersection of AI innovation and managed services excellence, helping MSP owners and operators build the AI practices that will define their businesses for the next decade.

Our Services

- **AI Revenue Architecture:** Full go-to-market strategy for MSP AI practices
- **Managed Intelligence Launch Program:** A complete AI services launch playbook
- **Revenue Launch™:** Accelerated revenue program for new AI service lines
- **Process Intelligence & Automation Readiness:** Captures workflows, centralizes processes, and governs execution, automation, and AI
- **Executive Retainer:** Ongoing strategic advisory for MSP CEOs and owners
- **Fractional CRO:** Specialized revenue leadership on a fractional basis

Connect with us:

aitadvisorygroup.com | info@aitadvisorygroup.com

We Put AI in IT™

www.ingramcontent.com/pod-product-compliance
Lightning Source LLC
LaVergne TN
LVHW080659011026
849160LV00014B/1481